Barbara Dorigo

CHLORELLA
Natural Medicinal Algae

by Dr. DAVID STEENBLOCK, B.S., M.Sc., D.O.
President, Aging Research Institute

Published by Aging Research Institute
El Toro, CA

I

Note to the reader:
Explanation of the front cover.

Center — Picture of Pre-Cambrian Chlorella fossil (X 3000) (2.5 Billion Years)

Lower Left — Dyno mill patented machine to break down the cell wall by more than 95%.

Upper Left — Flask seed culture in manufacturing process of Chlorella.

Upper Right — Main culture pool with stirrer. Photosynthesis with energy from sun.

Lower Right — Magnification of Chlorella cell (X 600).

Library of Congress Catalog Card Number: 87-070623
ISBN: 0-9618268-0-0

Printed in the United States of America

II

PREFACE

Prevention is the best "treatment" for disease. But when illness does occur, the next best treatment strengthens us sufficiently to throw off illness, rather than just the suppression of symptoms, adding nothing to our bodies' efforts to heal.

Proper nutrition is obviously among the best means of strengthening our bodies' defenses against disease. Although nutritional supplements are sometimes invaluable, whole foods are always the best sources of nutrients. Few whole foods have the unique nutrient-rich pattern found in chlorella, a "new" whole food from Japan.

In the pages which follow, Dr. David Steenblock has summarized for us chlorella's wide range of effects on human health, both preventive and therapeutic. He writes that it's not only excellent "basic nutrition", but has proven detoxifying effects, as well as effects against specific symptoms and illnesses.

Dr. Steenblock points out that chlorella is 60% protein, unlike nearly any other non-animal-source food. Chlorella is 20% complex carbohydrate, and only 11% fat (82% unsaturated). It's an excellent source of DNA and RNA, chlorophyll and Beta-carotene, as well as many other vitamins and minerals.

Vegetarians are constantly admonished by their doctors (including myself) to be especially careful about protein, iron, and vitamin B-12. Plant sources of B-12 are almost non-existent. Generally, plant foods contain less iron and protein than animal source foods. As an excellent source of protein and iron, and a "non-animal" source of vitamin B12, chlorella would appear to be an ideally complementary vegetarian food.

Dr. Steenblock has done an outstanding job gathering and presenting in understandable form much of the scientific information concerning this unique food substance. His book is a valuable "signpost" on the road to good health.

JONATHAN V. WRIGHT, M.D.

FOREWORD

I became interested in health food supplements when my personal medical problems couldn't be helped by conventional medical drugs and methods. During this difficult period of my life I traveled from one health food store to another searching out books, pamphlets, magazines and leaflets that would describe the various supplements and their medical uses so that I could decide if any would be beneficial to me. As a trained biochemist, physician and pathologist I was very much disappointed and frustrated in the lack of clear cut information about the various supplements and their medical uses. Often statements were made about the great benefits of this or that product but no scientific data would ever be included. Many of the health books and magazine articles would have scientific articles quoted but when I went to the library and read these quoted articles, the facts were not as the books and articles were claiming. This was another frustrating experience and if I had not been determined to regain my health in any way that I could, I probably would have quit my quest for knowledge of health supplements and may have concluded that it was all one giant "rip off". Unfortunately for me, at the time of my initial foray into health food stores, this small book on Chlorella had not been published nor was Chlorella available in this country. If Chlorella and this book had been available I am sure that I would have regained my health much quicker than I did and I would have not been nearly so frustrated with health food store books. With those personal frustrations behind me I decided that if I were ever to write a health food store type of book I would try to present the data as honestly as I could and to provide as many references as possible so that any enterprising person could go to the library and read the research papers themselves.

If you are a physician or health practitioner I would encourage you to study this book well and to try this material on your patients, realizing that its action is generally slow and that the person has to take it continuously for a matter of months before optimum results are produced. As of the spring of 1987 when this book was finished the scientific documentation of its various uses are in most part contained within this book. As you can see from the small number of pages and the large number of references, only the highlights of the scientific research have been included but I would estimate that it includes at least 98% of all the clinically useful information one needs in order to understand and to utilize this substance effectively. This book is intended for the intelligent layman and health practitioner looking for answers in regard to preventing and reversing health problems. I hope that this book will serve to educate the public of the benefits of chlorella and that the public discovers its usefulness so that each and everyone may live a long and healthy life.

DR. DAVID STEENBLOCK, D.O.
1987

TABLE OF CONTENTS

INTRODUCTION

Man's quest for health and long life has led to searching the far corners of the world for answers. This is the story of one such successful search. Chlorella could very well be the answer for many people who would like to maintain or regain their good health. Good health does not come accidently nor does it return quickly—especially if the problem(s) one faces are great. Chlorella is an all natural vegetable plant that, when taken over a long period of time has many beneficial effects on a variety of health problems.

As a practicing physician (general, preventive and restorative) it has been my privilege to clinically test this new product on a number of health problems. I have also reviewed the world literature on this important substance and am pleased to be able to present this information to the reader. A recent publication by Dyana Bewick reviews the history, manufacturing techniques and food uses of this substance and should be consulted by anyone interested in those aspects of Chlorella[1].

The following index lists most of the conditions shown to be improved by Chlorella and is presented in alphabetical order to help the reader quickly find the research that applies to his/her particular interest. The study of Chlorella's health benefits is just beginning and this book hopefully will be of use to physicians, laymen and scientists who are interested in further work with this unique substance.

My apologies to any research scientist or physician whose work was inadvertently left out. If you would be so kind as to send copies of your research I will include it in the next edition of this book.

A special thank you to my wonderful wife Noyemy for her patience, tolerance and love during this project.

I would also like to thank Lyn Darnell for proofreading and editing, Lyn McClure for her statistical work and the many scientists who did the actual research.

I hope you will find this book as exciting to read as it has been for me to investigate and write. Most importantly of all, take Chlorella and stay healthy!

WHAT IS CHLORELLA?

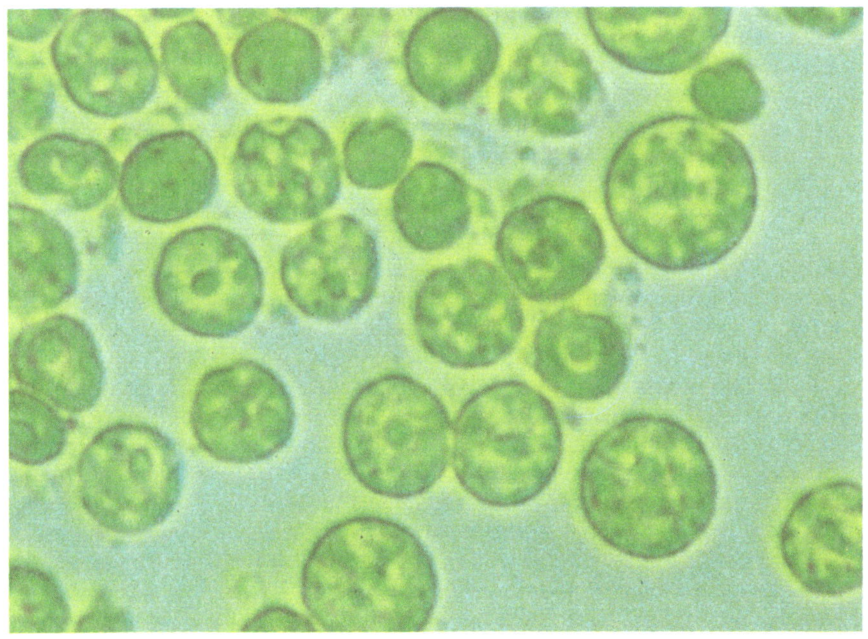

Chlorella is a green algae and appears above as seen through a microscope. Algae are a group of plants, one-celled, colonial, or many-celled, containing no true root, stem, or leaf. This particular fresh-water, single celled, microscopic plant contains a whole host of nutritious and health building nutrients. Chlorella gets its name from its high content of chlorophyll (the highest of any known plant). In addition to chlorophyll it contains vitamins, minerals, dietary fiber, nucleic acids, amino acids, enzymes, CGF (Chlorella Growth Factor) and other substances.

Chlorella has been on the earth since the Pre-Cambrian Period — over 2.5 BILLION YEARS! It was the first form of a cell with a true nucleus. Its survival to modern times is a tribute to its inherent genetic stability, hardiness and unusually effective DNA-repair mechanisms[3]. As we shall see, these characteristics are important to its healing effects in Man. Since it is the simplest form of a cell with a true nucleus, its study should give science a greater understanding of how cells function. Unfortunately, due to its small size, it was not until 1890 that Chlorella was discovered and named by Beyerinck[2].

Chlorella has a strong cell wall that prevents it from being adequately digested. For this reason, it was not until 1977 that it began to be used as a health food in the United States. With the discovery of a method (Dyno-mill™) to break the cell walls, the product became more digestible and easily tolerated. The Dyno-mill™ process has the ability to break down the cell wall by 95%. Other commercially available Chlorella uses a blanching, heat method to achieve the cell wall destruction and is more destructive to the cell's nutritional ingredients. Yet this cell wall is one of the main reasons that Chlorella is so unique. It helps the removal of hydrocarbon and metal poisons from the human body and helps stimulate the immune system. **Chlorella is the largest selling health food supplement in Japan** where the yearly production is 1,000 tons. After having reviewed the scientific literature it would appear that there are four main components of Chlorella which have been identified to have certain health effects. These are chlorophyll, the cell walls, vitamin A/beta-carotene and Chlorella Growth Factor (CGF). Chlorella has a number of properties which are helpful to organs and tissues that have been injured by a variety of causes. It is claimed to be a "great normalizer" changing high or low body functions back to normal.

Chlorella is marketed as pure tablets, powder, granules or as the water soluble extract which is called **CGF (Chlorella Growth Factor)**. The dosage of the tablets varies from 1 to 30 with each meal. The average preventive (maintenance) dose is 5 tablets with, after or between meals (15 - 20 per day). The person who decides to take Chlorella should not expect an overnight miracle since results usually start to become apparent after 3 months — however I have seen certain conditions improve very quickly, e.g. constipation and foul breath improve within one or two days while diabetic ulcers heal much faster with topical application of the CGF. Some people may feel improvement immediately and others may take up to 6 months or longer. The longer Chlorella is taken the more benefit should be enjoyed.

COMPOSITION

About 60% of Chlorella is protein, and about 20% consists of carbohydrates and fats. The proteins of Chlorella contain all the amino acids known to be essential for the nutrition of animals and human beings. In its amino acid composition, the quality of Chlorella protein can well be compared to animal proteins, the only drawback being its lower content of methionine. However, in certain cases (e.g. cancer) this may be an advantage since many cancers depend on methionine for growth. The vitamins found in Chlorella cells include: vitamin C, provitamin A, thiamine, riboflavin, pyridoxine, niacin, pantothenic acid, folic acid, vitamin B-12, biotin, choline, vitamin K, lipoic acid, inositol and para-aminobenzoic acid. The minerals include: phosphorus, potassium, magnesium, sulphur, iron, calcium, manganese, copper, zinc, and cobalt[164]. Such high nutrient content enables Chlorella to be of potential benefit to third world nations as a food source for their starving masses. In affluent countries on the other hand, it could be used as a high protein supplement for weight control as Dhyana Bewicke describes in her book "Chlorella: The Emerald Food"[1]

General Analysis

Moisture	3.6	%
Protein	60.5	%
Fat	11.0	%
Carbohydrate	20.1	%
Fiber	0.2	%
Ash	4.6	%
Calories	421	/100 g

Vitamins and Minerals

Vitamin A activity	55,500.0	IU/100g
B-carotene	180.8	mg/100g
Chlorophyll a	1,469.0	mg/100g
Chlorophyll b	613.0	mg/100g
Thiamine (vitamin B-1)	1.5	mg/100g
Riboflavin (vitamin B-2)	4.8	mg/100g
Pyridoxine (vitamin B-6)	1.7	mg/100g
Vitamin B-12	125.9	mcg/100g
Vitamin C	15.6	mg/100g
Vitamin E	< 1.0	IU/100g
Niacin	23.8	mg/100g
Pantothenic acid	1.3	mg/100g
Folic acid	26.9	mcg/100g
Biotin	191.6	mcg/100g
PABA	0.6	mg/100g
Inositol	165.0	mg/100g
Calcium	203.0	mg/100g
Phosphorus	989.0	mg/100g
Iodine	600.0	mcg/100g
Magnesium	315.0	mg/100g
Iron	167.0	mg/100g
Zinc	71.0	mg/100g
Copper	0.08	mg/100g

Amino Acids (expressed in w/w%)

Lysine	3.46
Histidine	1.29
Arginine	3.64
Aspartic acid	5.20
Threonine	2.70
Serine	2.78
Glutamic acid	6.29
Proline	2.93
Glycine	3.40
Alanine	4.80

Cystine 0.38
Valine 3.64
Methionine 1.45
Isoleucine 2.63
Leucine 5.26
Tyrosine 2.09
Phenylalanine 3.08
Ornithine 0.06
Tryptophan 0.59

Fatty Acids

Unsaturated fatty acids 81.8%
Saturated fatty acids 18.2%

C 14:0 . . . . .0.6%
C 14:1 . . . . .0.9%
C 14:2 . . . . .0.9%
C 16:0 . . . . .15.6%
C 16:1 . . . . .9.1%
C 16:2 . . . . .5.5%
C 16:3 . . . . .17.1%
C 18:0 . . . . .2.0%
C 18:1 . . . . .10.0%
C 18:2 . . . . .15.5%
C 18:3 . . . . .22.8%

Assay performed by Japan Dairy Technical Association, 3, Kioicho, Chiyodaku, Tokyo, Japan. Sample Chlorella G-Powder No. 650618. Date of Assay January 17, 1977. Inspection Number 1757.

The different Chlorella products available internationally have slightly different compositions depending on the strain of Chlorella used and the ingredients that are used as the raw materials. In addition, there is a difference in the types of products available. At least one company's product is the whole cell while other companies utilize cells which have had their cell walls broken down to increase digestibility.

The difference in total digestibility is:

Disrupted cells (Dyno-mill) 79.5%
Heated and blanched cells 50.0%
Whole cell Chlorella 47.0%

In vitro 5 hour pepsin digestion

CHLORELLA STIMULATES THE IMMUNE SYSTEM!

Probably the most exciting and potentially rewarding area of modern medicine is the field of immunology. **Immunology is the study of the immune system; the body's mechanisms developed to fight off foreign invaders, whether they be bacteria, viruses, chemicals or foreign proteins. The body's defenses have a unique way of inactivating or detoxifying each of these types of substances.**

B-cells are active in fighting against bacteria; **T-cells** are active against viruses and cancer; and macrophages are active against cancer, foreign proteins and chemicals. **Macrophages** are large cells that are located in tissues such as the liver (Kupffer cells), spleen, lymph nodes, thymus, lung, abdominal cavity, the blood (monocytes), joints (synovial lining cells), bone marrow and connective tissue which actively clean the blood, body fluids and cavities of harmful substances   (this process is called phagocytosis). Since there is a finite number of macrophages in a person's body, there is a limited ability to remove harmful substances from the blood. One of the ways used to fight cancer is the use of agents to stimulate macrophage production and activity. This macrophage stimulation causes increased cancer cell destruction and also increases the removal of harmful cancer debris from the blood by their phagocytic activity. **Interferon is a natural secretion of the body and is thought to be a physiological stimulator of macrophages**[85,86]. Injection of interferon into the body stimulates macrophages' oxidative metabolism and function[87]. When hot water extracts of Chlorella are injected into mice, high levels of interferon are observed in the blood 2.5 hours after the injection.

Fig. 3 Production of circulation IFN with Chlon A in mice

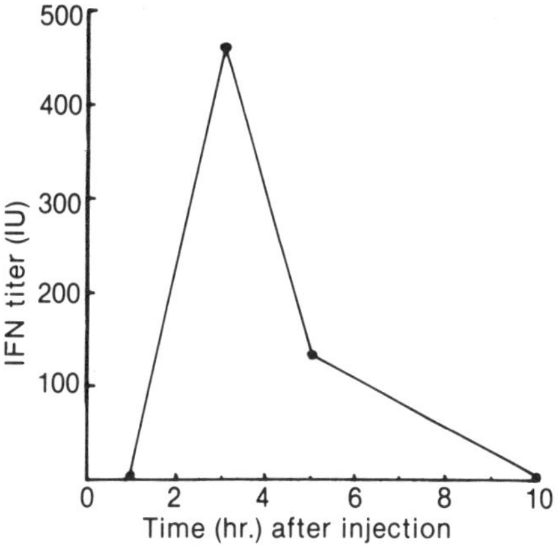

The first experiment to demonstrate the immune stimulating and detoxification power of Chlorella was performed in 1973 by Kojima and Associates[90]. These scientists injected a Chlorella extract into rats intravenously and 24 hours later injected particles of carbon into the same rats. They observed how fast these particles were removed from the blood.

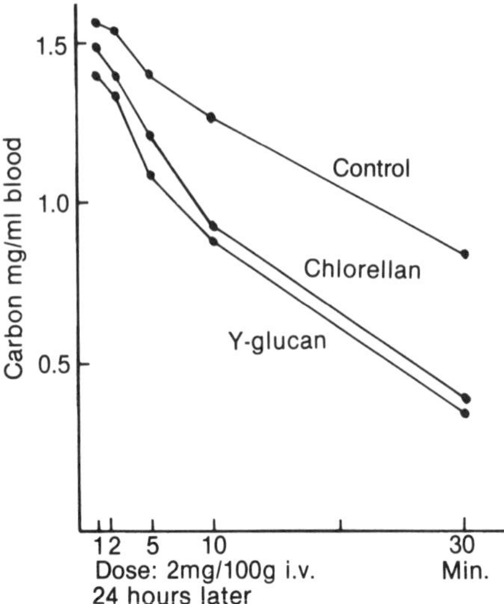

Carbon clearance of Chlorellan-treated rats and Yeast Glucan-treated rats. Each sample dissolved in 0.2ml of saline was injected intravenously at the rate of 2mg per 100g of rat body weight, and 24 hours later, carbon clearance test was carried out.

The maximal activity of the Chlorella extract was found to occur by 72 hours after injection and its stimulating activity slowly disappeared after two weeks.

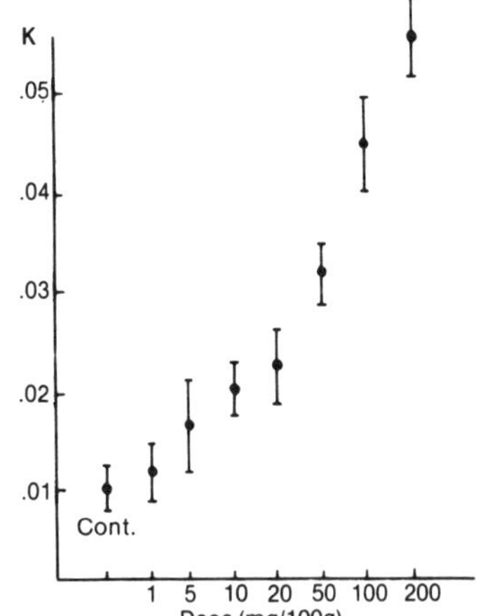

The relationship between accelerating activity and dosage of crude extract. Samples of crude extract, dissolved in 1 ml of saline, were intra-peritoneally injected and 5 days later a carbon clearance test was carried out.

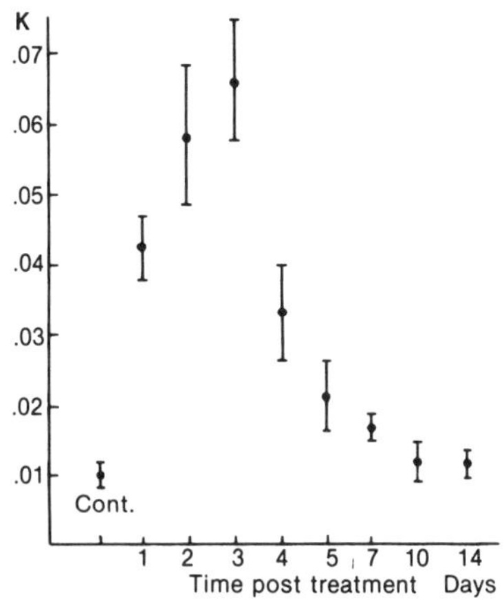

The relationship between accelerating activity of crude extract and period of the administration. Samples of crude extract, dissolved in 1 ml saline, were intra-peritoneally injected at the rate of 20 mg per 100 g rat body wt, and 1-14 days, carbon clearance was tested.

The material extracted and used in these experiments was shown to have a molecular weight of 1250, was water soluble, and was free of toxicity. The name "Chlorellan" was coined to describe this extract. Examination of the rat's tissues proved that there was increased activity of the macrophages. In a study to confirm this, macrophages from the abdomen were taken from rats previously injected with Chlorellan and examined under the microscope. These macrophages were much more active in taking up carbon particles as compared to the controls, indicating that activation of macrophage phagocytosis (the engulfing of microorganisms by the macrophage) had occurred. Further studies found unidentified serum factors that were also involved in stimulating the macrophage system[90]. Many more studies also demonstrate that Chlorella stimulates the immune system by way of macrophage stimulation[91-96].

CHLORELLA AND CANCER

Chlorella significantly increases the number of macrophages in tumor bearing mice — producing an anti-tumor effect and prolonging the lives of the animals[97].

Other recent studies have shown that another form of white blood cell, the **polymorphonuclear leukocyte, is activated by Chlorella extract in a nonspecific way to fight cancer cells**[99].

A recent study indicated that Chlorella derivatives (especially autoclaved cells and heat-extracted ones) enhance macrophage activity and cytotoxic activity of lymphocytes (possibly T-cells). The suggestion was made that the antitumor action of Chlorella was due to a synergistic effect of macrophages and cytotoxic lymphocytes[98].

The Chlorella cell walls have structures similar to the cell walls of bacteria (such as B.C.G.) and fungi that have also been shown to have anti-cancer effects due to their ability to stimulate the immune system[119]. Vermeil and Morin were the first to recognize this relationship and to prove that Chlorella cell wall material had an effect on cancer. They demonstrated that intra-peritoneal injections of Chlorella protected CH3 mice by 82% against sarcoma BP8 transplants[119]. Further work by that group indicated that the anti-tumor effect was possibly due to macrophage stimulation[120]. **An acidic polysaccharide prepared from Chlorella cell walls has been shown to induce interferon both in the test tube and in experimental mice.** The induced interferon was able to protect the experimental mice from infections with vaccinia and influenza viruses as well as showing anti-tumor activity against Ehrlich ascites carcinoma[101]. It thus appears that the polysaccharide components of Chlorella cell walls have effects on cancer due in part at least to its stimulation of interferon production. Interferon is a natural secretion of the body and is thought to be a physiological stimulator of macrophages[102, 103]. Injection of interferon enhances macrophages' oxidative metabolism and function[104].

The mechanism of immune stimulation by Chlorella would seem to be similar to the action of **interleukin I and II** on lymphocytes in that the substance activates the cells to become more active.

In treatment of cancer patients the number of certain lymphocytes and their relation to each other (helper/suppressor cell ratio) is important. Cancer patients often have depressed levels of helper cells and elevated levels of suppressor cells. **High doses of Chlorella have been shown to reverse this ratio in some patients (a decrease in suppressor cells and an increase in helpers).** These are the results seen in a patient with Hodgkins disease after 6 weeks of Chlorella (9 packets/day).

LYMPHOCYTE MARKER REPORT ....DATE..07-01-86

	RESULT	EXPECTED VALUES
TOTAL WHITE COUNT	4300	(4500-10300)
TOTAL LYMPHOCYTE COUNT	774	(1500-4000)
TOTAL T CELLS	163 (21%)	(800-2530) (65-79%)
SUPPRESSOR CELLS	77 (10%)	(220-865) (20-36%)
HELPER CELLS	101 (13%)	(480-1185) (35-55%)
NATURAL KILLER CELLS	39 (5%)	(45-650) (15-27%)
TOTAL B CELLS	85 (11%)	(60-400) (5-15%)
CYTOTOXIC CELLS	201 (26%)	($<$450) ($<$15%)
HELPER/SUPPRESSOR	1.30 /1	(1.65/1 - 2.3/1)

LYMPHOCYTE MARKER REPORT ....DATE..08-15-86

	RESULT	EXPECTED VALUES
TOTAL WHITE COUNT	5100	(4500-10300)
TOTAL LYMPHOCYTE COUNT	1326	(1500-4000)
TOTAL T CELLS	542 (41%)	(800-2530) (65-79%)
SUPPRESSOR CELLS	125 (9%)	(220-865) (20-36%)
HELPER CELLS	464 (35%)	(480-1185) (35-55%)
NATURAL KILLER CELLS	123 (9%)	(45-650) (15-27%)
TOTAL B CELLS	293 (22%)	(60-400) (5-15%)
CYTOTOXIC CELLS	149 (11%)	($<$450) ($<$15%)
HELPER/SUPPRESSOR	3.9 /1	(1.65/1 - 2.3/1)

The use of Chlorella in the prevention and treatment of cancer is still in the experimental stages and no one should elect to use Chlorella as a treatment for cancer except as an adjunctive aid when other appropriate and conventional therapies are also being used.

Most of the reported scientific reports have shown Chlorella to NOT have any DIRECT action against any cancer or tumor except for chlorophyll derivatives (see Photodynamic Therapy). What has been shown is that Chlorella when either given orally or by injection exerts a significant anti-cancer activity by stimulating the "host-mediated response system". This simply means that Chlorella stimulates the body's immune defenses so that the body fights the cancer better!

The anti-cancer activity appears to be present in both the Chlorella's cell wall (made up of acidic polysaccharides) and within the cells since its water soluble extracts have been shown to have anti-cancer activity as well[114].

As previously mentioned Chlorella contains chlorophyll. Chlorophyll has been studied as a treatment for cancer. When injected into test animals with induced and transplanted mammary carcinoma or sarcomas, porphyrins derived from chlorophyll, concentrate in the tumor (**porphyrins are chemicals derived from chlorophyll or hemoglobin which do not contain iron or magnesium**).

These substances also concentrate in incised or otherwise traumatized parts of the body, accumulating in the regenerative margins of the incisions. This shows that growing tissues and regenerating tissues in general have a definite affinity for these porphyrin-like compounds[115].

The possibilities of cancer treatment with chlorophyll have been investigated by Tixier and also by Dupont and Duhamel with encouraging results. Dupont and Duhamel injected chlorophyll derivatives (1-100 mg) intravenously or into the tumors of patients with cancer and found no adverse effects[116, 117]. I am aware of no further work using chlorophyll as a treatment for cancer in the medical literature.

PHOTODYNAMIC THERAPY OF CANCER

A key finding that led to one of our modern scientific approaches to cancer was an observation by A. Policard as long ago as 1924 who found that experimental tumors exhibited a spontaneous fluorescence when exposed to a Woods lamp (ultraviolet light). Nearly 20 years later Auler administered hematoporphyrin (a chlorophyll derivative) by injection to mice, and observed that it became concentrated in malignant (cancer) tissue. Lipson and co-workers in 1960 developed a derivative of hematoporphyrin which enhanced the amount of tumor localizing properties of the porphyrin. **After the porphyrin becomes concentrated within the tumor, exposure of the tumor to a certain wavelength of light will cause the release of singlet oxygen which kills the tumor by oxidizing fatty membranes[118].**

The degradation of chlorophyll yields products which are identical with the products of animal hemoglobin breakdown — hemopyrrole[64], etioporphyrin[65], phylloerythrin[66] and phylloporphyrin (similar to hematoporphyrin). Granick isolated two precursors of chlorophyll in Chlorella — protoporphyrin[67] and magnesium protoporphrin[68]. Protoporphyrin from Chlorella was demonstrated to be identical to the blood pigment heme[69, 70]. The current drug used for cancer photodynamic therapy is dihematoporphyrin ether/ester which is similar to the hematoporphyrin found in Chlorella[173, 174]. These Chlorophyll-derived substances are absorbed into the body since they are water soluble and cross through membranes.

Another Chlorophyll-derived porphyrin called **pheophorbide** has been identified in Chlorella in small quantities[178, 180, 181]. In large amounts it has caused photosensitivity reactions in the skin of people exposed to sunlight[175]. These cases of photosensitivity occurred during a period from April 1976 to April 1977 and were caused by Kenbi Chlorella tablets from Taiwan. It was discovered that the company had altered its manufacturing process during this time and it was concluded that the manufacturing process was at fault since their dried Chlorella cells per se did not cause a reaction[182, 183].

The skin injuries consisted of swelling of the endothelial cells and thrombosis of small blood vessels in the dermis and the subcutaneous fatty tissue. This is similar to the changes produced in cancerous tumors when treated by photodynamic therapy. Other studies have demonstrated that the damage is due to the production of singlet oxygen (a type of oxygen with a free electron) and to its damaging effect on the fat contained within cell membranes (lipid peroxidation). **Of great interest is that these scientists demonstrated that the Anti-oxidants vitamin E, coenzyme Q, vitamin C, beta carotene and vitamin B-5 (pantothenate) all protected against this type of damage[176, 177, 179]. This brings up the possibility that those persons who developed photosensitivity reactions were low in these various anti-oxidants.**

Photodynamic therapy is one of the more exciting, effective and safe treatments of cancerous tumors currently available. Its only drawbacks are that the tumor must be accessible to a high intensity light source and that the person's skin may become sensitive to the sun for a few weeks. No studies have been done as of yet to test whether addition of these anti-oxidants to a person's diet after photodynamic therapy will reduce or eliminate this photosensitivity. This is a study that should be done very soon in order to make this new type of cancer therapy safer. While pheophorbide, Chlorella cell wall material and chlorophyll may have significant properties in regard to tumors, Chlorella cell extracts (CGF) also appear to have beneficial activities.

CHLORELLA EXTRACT

The hot water extract of Chlorella has been called Chlorella extract solution, Chlorella extract, Chlorella growth factor (CGF), Chlorella growth extract (CGE), Fujimaki material, F.S., and Controlled growth factor. Most authorities and popular usage, favor the term Chlorella Growth Factor (CGF). Estimates of the content of CGF contained in raw Chlorella are usually around 5%[122].

"The real benefits which fresh water microalgae can provide to modern man, when eaten on a daily basis in one to three gram amounts, is not in its protein food value, but instead is in a "Controlled Growth Factor"—designated as CGF by Japanese biochemists to describe the combination of molecules that provide the large increase in sustaining energy when certain types of algae are eaten by man..." Kolman, H.V. and R. Schmidt [123].

CGF is not a single substance but contains a variety of substances such as amino acids, peptides (like glutathione), proteins, vitamins, sugars, and nucleic acid — related substances such as adenosine nucleotide and cytidine nucleotide. A glycoprotein extracted from Chlorella was found to have an anti-tumor activity against sarcoma 180 (a type of cancer) implanted in mice and also against mouse leukemia in culture[124].

Giving by mouth an extract of Chlorella to mice that had been inoculated with sarcoma 180   inhibited tumor growth by 52.9% at the end of 25 days of tumor growth. These same scientists reported successful results against rat ascitic liver cancer[125].

Hot water extracts of Chlorella significantly inhibited the growth of Methylchloanthrene induced fibrosarcomas in mice when the extracts were injected into the tumor or into the surrounding tissue[126].

Other studies have shown Chlorella to be active against **breast cancer**, and **liver cancer** in mice when given orally or injected into the abdomen[127].

Recently, preliminary human clinical trials have been conducted with injections into the abdomen showing a 'good tolerance'[128]. Professor C. Vermeil of the University of Nantes in Nantes, France stated in a letter to me that **"...the utilization of the chlorellas is most interesting and promising as an anti-cancerous agent administered into the peritoneum. It is very easy to cultivate these algae and they are totally harmless to the human peritoneum."** Nine patients with cancers of the female organs, colon and liver received one to five injections weekly of 2 grams of peptido-glycans prepared from Chlorella pyrenoidosa (the genus species sold in health food stores). Subjective improvements were noted in correction of weakness and loss of strength and 2 patients experienced a remission but other treatments were being given so no conclusions can be made about its effectiveness[79].

Recently, cancer cells have been shown to invade normal tissue by the use of certain enzymes (endoglycosidase, cathepsin B, plasminogen activator and most importantly type IV collagenase)(129). This is a fundamental process which enables cancer cells to spread through the body and leads to the death of the individual. Usually the original cancer is not what kills a person but the small 'seeds' of cancer which spread and invade other tissues and grow. Since Chlorophyll has been shown to have anti-proteolytic (anti-protein digesting enzyme) activities, Chlorella with its high amount of chlorophyll may have an important role to play in the prevention of the spread of cancer (metastasis) by inhibiting these cancer cells' protein-digesting enzymes.

That Chlorella may help the body fight cancer is beginning to be shown by way of scientific studies of vitamin A and beta-carotene's cancer preventing aspects. Due to the fact that there are literally thousands of papers that have studied vitamin A and beta carotene and their effects on the immune system and cancer prevention and treatment, space will not allow a complete review of these studies. In general, vitamin A affects immunity by: 1) maintaining the mechanical integrity of tissues and mucousal surfaces such as the linings of the mouth, esophagous, intestine and skin; 2) acting as a nonspecific enhancer; and 3) stimulating the body's phagocytes to attack foreign invaders and cancer cells. People who develop cancer have been shown to often have lower levels of vitamin A in their blood than other people. **People who eat lots of green and yellow vegetables have less cancer due to the high vitamin A content contained in their diet. In addition, supplementing the diet with vitamin A has also been shown to help prevent cancer**(170).

When high doses of vitamin A are combined with radiation therapy in the treatment of cancer, the treatment dose of radiation can be reduced by 50% and the person can tolerate larger doses with less side effects(130, 131, 132).

The Robert-Janker-Klinik in Bonn, Germany has pioneered the use of high dose vitamin A in the treatment of cancer. In 1971, this group published the results of 37 cases of invasive cancer of the vulva, treated with high dose vitamin A and radiation. The amount of radiation was reduced by 1/3 in comparison to those receiving only radiation. Similar results were seen in esophageal cancer. Dr. Wolfgang Scheef of the Janker Klinik gave his results of the past 20 years using high doses of emulsified vitamin A in humans at the recent 14th International Cancer Congress in Budapest (1986). He reported almost an 100% cure of low grade squamous cell carcinomas of the skin as well as significantly beneficial results in a number of other types of cancer(133) with high doses of vitamin A.

BETA-CAROTENE KILLS CANCER CELLS

Chlorella contains 180 milligrams of Beta-carotene per 100 grams. **Beta-carotene has been shown to destroy cancer cells and enhance macrophage production of tumor necrosis factor (TNF)** and the T-helper cell stimulator, interleukin I. **Beta-carotene works synergistically with vitamin E as an anti-oxidant to squelch cancer in its initial stages.** Drs. Joel Schwartz, Diana Suda and Gerald Shklar of the Harvard School of Dental Medicine reported this research at the 1986 meeting of the American Academy of Oral Pathology in Toronto. They showed a dose response effect of beta-carotene on hamster cheek cancers induced by the carcinogen 7, 12-dimethylbenzanthracene. **Extracts of algae were also studied and were shown to be more effective than just Beta-carotene,** causing the Harvard group to speculate that other factors are to be found in algae which give algae more anti-tumor effects than can be accounted for from the Beta-carotene content alone[100].

Cancer is a disease process in which new cells are constantly forming but also older cells are constantly dying. It is the secretion of the dead cell wall materials from the growing cancer mass which poisons the immune system of the body, making it incapable of fighting back. By speeding up the removal of these substances from the body, the immune system becomes more effective in fighting the cancer. The effect of Chlorella on stimulating the liver and other detoxification mechanisms has not been studied in relation to metastatic cancer but the potential for it to work in helping clear the blood of cancer cell debris would seem to be likely, since it does stimulate the removal of particulate matter from the blood[90, 95, 96].

When cancer victims take chemotherapy, radiation or various medications, their stomachs and intestinal tracts often become disturbed. Since Chlorella has soothing and healing properties on the stomach, I suggest taking it in liquid form. Use of the granules or powder is best and may be taken in distilled water either by itself or mixed in a blender with other health-building ingredients. The taste is semi-unpleasant so that the addition of other ingredients will not generally make the taste worse. Fresh carrot juice may be used in place of water for additional immune stimulating properties. If loss of weight or weakness is a problem the Chlorella and carrot juice mixture could have the whites of 1-2 eggs and/or dessicated baby beef liver powder (1-2 tablespoons) all combined in a blender. This recipe combines what the body needs to help the liver and detoxification mechanisms work more effectively in removing cancer cell wall debris from the blood. In addition, it has plenty of protein and immune stimulating properties. Some people mix this with tomato juice to cover the taste.

THE REVIVER

Chlorella 2 - 4 packets of granules, or 1 heaping tsp powder or 30 - 60 tablets

Fresh carrot juice 4 ounces

Dessicated baby beef liver powder 2 tablespoons

Egg whites from 2 eggs.

1 capful of the CGF

Mix and drink three or more times per day.

Immediately wash out your mouth and suck on a half of an orange to kill the taste.

This mixture contains high quantities of vitamin A and should not be taken for prolonged periods (more than 1-2 months) without physician supervision.

If the intestines don't seem to like this mixture or are severely disturbed with abdominal gas or pain, add 1 to 2 teaspoons of one of the high potency acidophilus preparations to the mixture. If unsure of the state of the intestine, look at the tongue. If the tongue is coated, add the acidophilus. If still having problems, add 1 tablespoon of aloe vera juice to the mixture. If still unable to take the mix, then decrease the amounts of everything by 2/3 and start over again.

CHLOROPHYLL

In addition to CGF, Chlorella also contains more chlorophyll per gram than any other plant. Comparison with spirulina shows that spirulina contains 7.60 gram per kilogram of chlorophyll while **Chlorella contains four times** as much (28.9 gram per kilogram). The 7.6 gm/kg figure for the amount of chlorophyll in spirulina is found in Christopher Hill's book "The Secrets of Spirulina" which also contains a ten page glowing report on the positive effects of chlorophyll which "Spirulina contains in abundance" [p.171-181].

Chlorophyll is an interesting substance in that it has found application in industry for coloring candles, waxes, resins, fats, oils, confectionary, gelatin, egg white, chewing gum and in general many food materials[4].

Chlorophyll consists of two components: chlorophyll a and chlorophyll b which exists in a 2.9 to 1 ratio in most plants.

Since chlorophyll's chemical structure is similar to hemoglobin's (the red pigment of blood), several scientists have suggested the use of chlorophyll as a medical therapy for anemia. A complete review of chlorophyll and its effectiveness in correcting anemia as well as other medical uses was presented by Kephart[13] in 1955. Briefly, if the person is not lacking iron or copper, the addition of chlorophyll may stimulate the production of blood presumably by providing the precursors to hemoglobin.

Much research has been done which has proven that **chlorophyll or its derivatives** influence bacterial and animal growth, metabolism, and respiration. Chlorophyll or its derivatives **stimulate the formation of erythrocytes in the blood**[11, 12], and **affect nutrition, synthesis of vitamins in plants, hormone action, tumors, and a number of diseases, like anemia, arteriosclerosis, cardiac hypertension and others.**

The pioneering work to introduce chlorophyll into medicine was done by Burgi, who used chlorophyll preparations under the name of Chlorosan[5] for wound healing.

Wound Healing Properties and Tissue Regeneration

In 1930 Rollet and Burgi first demonstrated that **extracts of green plants had stimulating effects on tissues growth.**[6].

By 1932 there was enough medical research done on chlorophyll that Burgi was able to publish a review monograph[7].

Gordonoff and Ludwig in 1935 showed that extracts of plant pigments have a stimulating effect on the growth of tissue cultures[8].

In 1937, in experiments with artificially inflicted wounds on the skin around the spinal column of rabbits and ginea pigs, Burgi was able to show that **pure chlorophyll has far greater stimulating power over the regeneration of tissue than either carotene or xanthophyll** (a yellow colored pigment of plants, i.e. oxygenated carotenoids)[9].

In 1943, 3 physicians at the New York Post-Graduate Medical School studied chlorophyll in the treatment of a variety of types of **skin ulcers** and found that 19 of the 25 ulcers studied responded favorably. The chlorophyll was used in ointment form or as a 0.2 percent solution and appeared to have a "stimulating effect" on the supportive tissues promoting rapid healing[10].

In a very extensive study, Smith and Livingston studied the healing of 1,372 experimentally induced wounds and burns by the topical application of seventeen medicinal preparations. Of all agents studied only chlorophyll preparations showed consistent statistically significant positive results. **Wound healing was accelerated by 24.9% and occurred in 67.9% of the cases.** The only wounds that did not show improvement with the chlorophyll preparations were infected. On the basis of their investigations they suggested that chlorophyll preparations "be used much more extensively in the treatment of wounds and burns"[17].

After determining conclusively the power of chlorophyll's healing abilities Dr. Smith went on to study infected ulcers with chlorophyll by itself and in combination with various antibiotics. When chlorophyll was used in combinations with penicillin in infected ulcers, the healing rates of wounds and burns were accelerated by 35% over the use of chlorophyll or penicillin by themselves[18].

In another study Dr. Smith demonstrated that chlorophyll in concentrations between 0.05 to 0.5% in tissue culture caused an **"almost immediate growth response" of fibroblasts** — the cells the body uses to repair wounds[24].

Gruskin gave a review of 1200 cases benefited or cured by application of chlorophyll solutions. These included acute infections of the upper respiratory tract and sinuses, and chronic ulcerative lesions of various types.

Salt Water Nasal Wash for Sinusitis and Nasal Congestion

Add 1/2 tsp. table salt to 1 pint lukewarm water

1) Place solution in soupbowl.
2) Over sink, block one nostril from side with index finger.
3) Dip nose into solution and inhale through nose to bring liquid into mouth.
4) Remove bowl from under nose and allow solution to drain from nose and mouth.
5) Repeat until 1/2 solution is used per nostril.

This standard remedy for congested sinuses and nostrils may be modified by boiling Chlorella granules (1 packet granules or 1 tsp powder per pint water), strain through cheese cloth while hot and then use as described above (i.e., add the salt to this pint of water + Chlorella).

Dr. Gruskin reported chlorophyll's usefullness in **ulcerative carcinoma** where a great deal of putrefaction with associated foul odor is present. Dr. Gruskin at the time of this article was the Director of Experimental Pathology and Oncology at Temple University. He suggested that **"the action of chlorophyll consists for the most part of increasing the resistance of cells in some physicochemical manner so that enzymic digestion of the cell membrane by invading bacteria or their toxins is checked."** This suggestion would seem to be the first mention of it as an agent that inhibits cell destruction caused by enzymes (antiprotease activity; see Pancreatitis, p. 19). Another suggestion was that it is effective in removal of odors caused by anaerobic (non-oxygen requiring) bacteria due to its ability to break down carbon dioxide and set free oxygen. Dr. Gruskin also reported on the non-toxic nature of chlorophyll when given by mouth and even when given intravenously[14].

Bowers discussed the extensive healing effects of chlorophyll on infections of the ear, nose and throat[15].

Pyorrhea (bleeding of the gums and loose teeth) was reported to be improved by chlorophyll, in a report by Goldberg in which he studied 300 patients. **Vincents stomatitis** is an infection of the gums which occurs in persons under stress, who have a vitamin C deficiency or practice poor oral hygiene. Dr. Goldberg reports "In Vincent's stomatitis chlorophyll has regularly brought about complete recovery, and much more promptly than with other agents." The technique was to spray the mouth with chlorophyll around and in between each tooth, at least daily. By means of an eye dropper the chlorophyll is squirted between the teeth three or four times per day. The same thing could easily be done with Chlorella by purchasing the granules or powder, dissolving in water and using the solution as described for chlorophyll. I would recommend the use of a water-pic type of instrument to wash the liquid Chlorella into all of the cracks and crevices of the gums for treatment of either pyorrhea or gum disease.

"In pyorrhea the use of chlorophyll has resulted in the tightening of teeth, the cessation of bleeding from the gums, and has grown new tissue."[16].

A more complete review of chlorophyll's effectiveness in treatment of gingivitis and pyorrhea can be found in Kemphart's excellent review[13].

Topical application of Chlorella extract has been found to be effective on cases of **granulomas, bedsores,** and **ulcerations of the toes**[19].

The usefulness of chlorophyll[20] and Chlorella in stopping odors and inflammation has even been used to treat **cervicitis** — the inflammation of the uterine (womb) cervix. Chlorella was inserted high in the vagina in a series of 100 women after each had their uterine cervix frozen as a treatment for cervical disease[21]. Significant results were obtained with the Chlorella treatment.

Radiation Burns

A topical ointment of chlorophyll was shown to be effective in treatment of the inflammation that occurs in the skin after radiation treatments for cancer and other conditions[22].

DIABETIC ULCERS

The healing of skin ulcers depends in large measure on the function and activity of a special type of cell — the fibroblast. Chlorella is a stimulator of interferon production. Interferon stimulates macrophages (as does Chlorella) and has been shown to stimulate the secretion of tumor necrosis factor (TNF)[35].

TNF has recently been proven to be a growth promoting factor for fibroblasts and this may be one of the reasons that Chlorella promotes the healing of skin ulcers so rapidly[36].

Diabetic ulcers are often very difficult to heal. If these ulcers become infected, as indicated initially by a redness around the area, the problem can get worse in a hurry (within one or two days) and the person's foot may have to be amputated or even life may be lost due to the infection. Because of the seriousness of this, anyone having this type of problem should be under the close supervision of a physician. If the wound is infected, the person needs to be given either oral or intravenous antibiotics and closely supervised. Chlorella Growth Factor in Honey can be used directly from the bottle for treatment of this type of ulcers. These pictures illustrate the healing power of Chlorella Growth Factor (CGF) in honey for treatment of a diabetic ulcer. CGF in honey can be applied directly to the ulcer in generous amounts and held in place by the use of a non-sticking type of dressing over which are applied 4 × 4 gauze pads soaked in the CGF. The purpose is to keep the ulcer saturated at all times with the CGF in honey. The person should keep entirely off their foot until healing is complete. A successful outcome is possible if the person carefully follows their physician's directions.

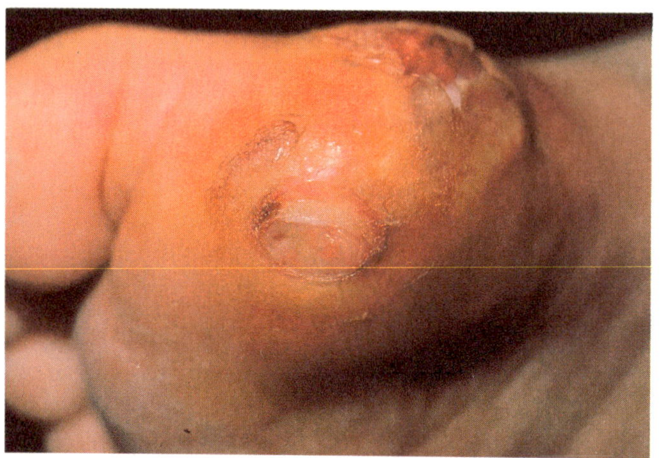

Before treatment the foot is grossly infected and ulcerated.

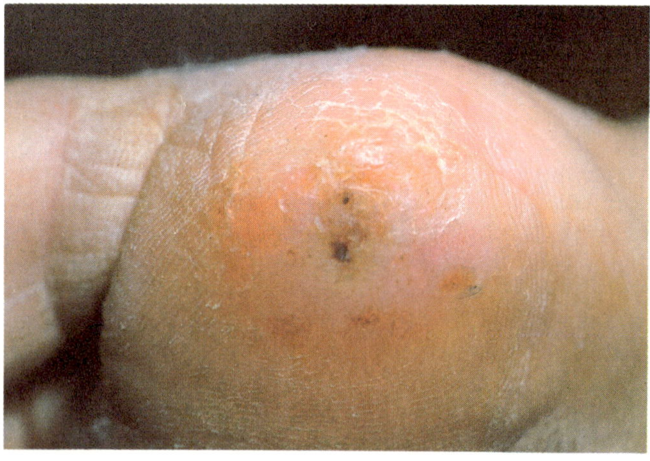

After treatment with Chlorella honey extract and other medicaments. By permission of the Rev. Walter Martin.

PANCREATITIS

Pancreatitis is an inflammation of the pancreas that can be very painful and life threatening. It may be caused by excess consumption of alcohol, by gallstones, poor nutrition, excessive calcium in the blood, and excessive blood fat. In a few cases no cause can be determined[37].

Chlorophyll, present in high concentrations in Chlorella, has been shown to be helpful in the treatment of pancreatitis.[38, 39, 40, 41, 42].

The treatment of pancreatitis with chlorophyll has been reported in a number of scientific papers most of which are written in Japanese.

The mechanism of how chlorophyll inhibits pancreatitis is of note. Yoshida, et. al., have demonstrated that **chlorophyll-a derivatives inhibit enzymes that break down proteins (proteases) — these enzymes are involved with inflammation and are in large part responsible for the harmful effects of pancreatitis**[43, 44].

Chlorophyll has been shown to inhibit the protein digesting enzymes — *trypsin, chymotrypsin, kallikrein, alpha-amylase, and phospholipase A* — in these various studies[45-57].

Trasylol has been one of the standard remedies used for pancreatitis since it inhibits trypsin — an enzyme that is elevated in pancreatitis. It may, however, produce allergic reactions. On the other hand, chlorophyll was found to have no side effects even after 10 years of studies in humans with pancreatitis. Of further interest is that recent randomized clinical trials have failed to show that trasylol is effective in acute pancreatitis which should serve to stimulate further investigation of the use of chlorophyll or Chlorella for the treatment of pancreatitis[58].

In addition chlorophyll was shown to be fairly effective in treatment of chronic pancreatitis in a human study published in 1980. Chlorophyll was used as an intravenous water soluble preparation on 34 cases with fairly effective results in 23 cases and some favorable effect in 9 others[59, 60].

Chlorophyll is metabolized by the liver and excreted in the bile[61, 62, 63].

The degradation of chlorophyll yields products which are identical with the products of animal hemoglobin breakdown — hemopyrrole[64], etioporphyrin[65], and phylloerythrin[66]. Granick isolated two precursors of chlorophyll in Chlorella — protoporphyrin[67] and magnesium protoporphrin[68]. Protoporphyrin from Chlorella was demonstrated to be identical to the blood pigment **heme**[69, 70]. Chlorophyll derived **heme** pigments have been shown to be absorbed into the body (the proof being their ability to correct some forms of anemia). Heme is an essential component of the cytochrome P-450 detoxification enzyme system (the main detoxification system of the body), and with Chlorella's high content of chlorophyll, taking Chlorella and its heme-like substances should help keep this system working effectively.

CHLORELLA AND DETOXIFICATION

Detoxification refers to the removal of toxic substances from the body. These substances are either poisons that have entered from the outside of the person such as occurs in pesticide poisoning or may occur from within the body, e.g., when the colon contains bacteria that produce toxic substances or as a result of inefficient metabolism of the body as a whole.

The detoxification capability of Chlorella is due to its unique cell wall and the material associated with it. The cell walls of Chlorella have been shown to have three layers of which the thicker middle layer contains cellulose microfibrils[73].

Atkinson, et. al., found a 14 nm-thick trilaminar layer outside the cell wall proper which was extremely resistant to breakage and thought to be composed of a polymerized carotene-like material[74].

The general analysis of the cell wall has shown an approximate composition of 27% protein, 9.2% lipid, 15.4% alpha-cellulose, 31% hemicellulose, 3.3% glucosamine, and 5.2% ash (containing iron and calcium)[75].

Cadmium Poisoning Reversed

Chlorella binds strongly to cadmium and will not give it up to the body. A study was done in which rats were given Chlorella that contained cadmium to determine if the cadmium would be absorbed from the Chlorella into the rats. In rats given only cadmium (no Chlorella) for 10 days, growth retardation was noted while no problem with growth was seen in those given Chlorella containing cadmium. Blood levels of cadmium were determined and also demonstrated that the cadmium that was bound to the Chlorella was not absorbed into the rat's body[184].

The clinical usefulness of Chlorella to detoxify cadmium was shown by a scientific study by Hagino, et.al.,[71] who demonstrated increased excretion of cadmium in persons suffering from "Itai-itai" (cadmium poisoning).

Dr. Ichimura reported that he gave 8 grams daily of Chlorella to patients suffering from pain caused by cadmium poisoning and obtained good results. **When Chlorella was given for 12 days, cadmium in the excretions of the patients increased 3 times over the baseline. After 24 days of Chlorella the cadmium in the urine had increased 7 times greater than baseline** and the pain of the patients was markedly reduced.

Pesticides, Insecticides, P.C.B. Removed

Chlorella has been used to detoxify people suffering from **P.C.B.** exposure (**polychlorbiphenyl**). Doctor Ueda of the Kitakyushu City Institute for Environmental Pollution Research in Japan gave 30 patients who suffered from P.C.B. exposure, daily doses of 4-6 grams of Chlorella for a one year period. As a result almost all improved (less tired, better digestion, normal bowel movements).

Besides P.C.B., **chlordecone (kepone)**, another very harmful chlorinated hydrocarbon insecticide, **has been shown to be removed more than twice as fast from the body** when Chlorella is taken by mouth. Dr. Pore of the School of Medicine, West Virginia University did a study in which Chlorella given to rats speeded up the detoxification of this toxin, decreasing the half-life of the toxin from 40 days to 19 days. The ingested algae passed through the gastrointestinal tract unharmed, interrupted the enteric recirculation of the persistent insecticide, and subsequently eliminated the bound chlordecone with the feces. Laboratory studies showed that there were two active absorbing substances — **sporopollenin** (a naturally occurring carotene-like polymer which is resistant to degradation) and the algae cell walls. Chlorella vulgaris contains no sporopollenin yet changed the half-life from 40 to 32.7 days, demonstrating that other forms of Chlorella have some detoxification properties even without sporopollenin. The commercially available products contain Chlorella pyrenoidosa which does contain sporopollenin.

This property of detoxifying hydrocarbon pesticides and insecticides is very important in our world of constant chemical exposure and is one of the important differences between Chlorella and other "green" products[72].

Another example of Chlorella's detoxification powers is a study in which a Brewer's yeast culture was poisoned and killed by the addition of **P.C.B., mercury, copper and cadmium.** When Chlorella extract was added to these toxic substances, the brewer's yeast remained alive! Chlorella has the unique property of picking up toxic substances from the surrounding environment (even in your body) and holding onto them. Other studies have shown that Chlorella cell walls absorb and hang onto **uranium**[76]. and **lead**[172].

Liver Health

Fink studied Chlorella as a food in rats and concluded that it **prevented liver gangrene** in these animals. The agents that were thought to be exerting these effects were thioamine acid, vitamin E, and an unidentified substance (factor 3 of Schwarz). Fink suggested that people suffering from kwashiorkor (a syndrome produced by severe protein deficiency) should be given protein made from Chlorella rather than that from milk[77].

Chlorella has also been shown to **protect the liver from toxic injury** due to ethionine. Ethionine is a drug which induces a fatty liver type of injury much like the liver damage that malnutrition produces. When rats maintained on a semisynthetic diet low in protein and vitamins were given ethionine, fatty livers developed. When Chlorella was added (5% of basal diet) the rats suffered less liver injury and recovered more rapidly from the poison[78]. Other studies have shown elevated levels of albumin and decreased levels of globulins while taking Chlorella which is what you would expect to see if the liver is working better.

Alcohol Hangovers Prevented?

Detoxification doesn't stop at the intestine with Chlorella. Professor Fukui of Shapporo Medical University reported in his book on Chlorella, experiments showing the effect of Chlorella in the detoxification process. Dr. Fukui reported that even with a fairly large consumption of alcohol, hangovers can be prevented up to 96% with the taking of 4 - 5 grams of Chlorella before drinking. The **liver's detoxification abilities are enhanced** by the Chlorella — promoting the removal of the alcohol by the liver[79].

Bowel Toxicity

As more research is done many diseases are being discovered that are aggravated by poor bowel health.

Bowel toxicity is difficult to measure in the human but can be done. Although not very objective or scientific, smelling a person's breath is one way of determining if a person is suffering from bowel toxicity. The problem with this method is that there are many other conditions which cause bad breath. Checking the urine for indican is also a method that will give some indication as to whether a person suffers from bowel toxicity. Chlorella helps clear bowel toxicity within a few days.

Why does Chlorella detoxify the bowel?

As has been mentioned, Chlorella contains more chlorophyll per gram than any other known plant. Chlorophyll has a long history of being effective at deodorizing bad smells — remember chlorets the gum for bad breath and the cat litter containing chlorophyll? Chlorophyll has been used and documented to act as an **underarm deodorant** and to **control bad breath**[25, 26, 27, 28, 29]. "It is common knowledge among workers in nursing homes, geriatric hospitals and mental institutions that chlorophyllin is an important aid in the **control of odors from incontinent patients."**[30]. As long ago as 1951, Weingarten and Payson reported that daily administration of water-soluble chlorophyllin **markedly reduced the odor in patients with colostomies**[31].

In 1944, Smith reported on the anti-bacterial properties of water soluble chlorophyll derivatives. As a result of these investigations, which were supported by a grant from the Committee on Therapeutic Research of the Council of Pharmacy of the American Medical Association, Smith concluded that chlorophyll acts to produce an environment unfavorable to the growth of bacteria instead of by any direct action upon the organisms themselves. The ability of these derivatives to hold back anaerobic bacterial growth points, in his opinion, to an oxidation mechanism. In brief, chlorophyll is not a killer of bacteria but does have a definite anti-bacterial growth action and may even display a bacteriocidal effect against certain organisms under suitable conditions[81].

In addition to chlorophyll's action on anaerobic bacteria, Chlorella's cell walls act to absorb poisons within the intestine and promote normal peristalsis. The intestinal tract (especially the small intestine) is lined by patches of lymphocytes which are probably stimulated by the Chlorella cell wall material to increase their abilities to destroy foreign invaders such as anaerobic bacteria. Chlorella also stimulates the growth of the good intestinal bacteria (lactobacillus) and those that produce vitamin B-12[80].

The stimulating and detoxifying effect of Chlorella on the bowel produces an interesting result at times. Many people will release more gases than usual for 3 to 7 days after beginning to use it. It is believed that the harmful intestinal bacteria are being fermented and destroyed. After this initial adjustment period, the bowel functions better than before and the gas problem disappears[82].

The deodorant properties of water soluble chlorophylls are dependent on the acid-base (pH) balance of the material that needs to be neutralized. Chlorophyll derivatives exhibit deodorant actions on the neutral or alkaline side with a pH optimum between 8 and 10.5. At the same pH chlorophyll derivatives exhibit antibacterial properties. It is of great clinical interest that bowel toxicity and constipation are associated with a stool pH of greater than 7, with the worst cases being of pH 9. Chlorophyll and Chlorella thus would seem to function best on treatment of colons that are suffering from anaerobic bacterial overgrowth syndromes — usually associated with a stool pH of greater than 7. Anyone can test their stool pH by buying pH paper from their local drugstore, taking a small amount of stool in a small amount of water, mixing well and inserting the pH paper into the solution and observing the color change[33].

Chlorella could be used for control of odors in the same manner that chlorophyll has been used due to its high chlorophyll content.

Constipation

Chronic constipation is another one of those problems that seem to afflict many of us as we grow older. Besides infrequent and hard stools, constipation may cause bad breath, headaches and has been associated with the development of pancreatitis, hypoglycemia and even cancer (breast and colon).

Dr. Young and Beregi reported that chlorophyllin tablets given in sufficient quantity would usually relieve the **chronic constipation** problems that nursing home patients often suffer. **Relief of intestinal gas** in terms of amount and odor was also observed with this type of chlorophyll in 85% of the cases[32].

Chlorella cell walls stimulate the intestine's linings and if taken for a few months will gradually strengthen the intestine, thus eliminating constipation. Chlorella also alters the bowel's content of bacteria by increasing the number of certain healthy bacteria. In one study five men between the ages of 24 and 30 years were given a control diet for 15 days.

23

Then for 30 days they were given a diet with only vegetable protein of which Chlorella was 50% followed by 10 days with the control diet. While the Chlorella diet was eaten — the health promoting bacteria lactobacilli, streptococci and bifidobacteria in intestinal microflora increased in comparison with control values[83]. (These streptococci are not the same as those that give a person a strep throat but instead promote the health of the bowel).

It has been demonstrated by many clinical cases that the administration of Chlorella algae is effective for treatment of constipation. Saito[84] thought of applying the promotive effect of Chlorella on intestinal peristalsis to treat persistent abdominal gas. Increased abdominal gas often occurs in cases of severe spinal cord injury where there is complete paralysis of the lower half of the body. These are a few of the cases he studied.

CASE : A 48 year old woman with a compression fracture of the third lumbar vertebra complicated by severe injury of the spinal cord. She had complete paralysis from the abdomen down. Persistent abdominal gas occurred soon after the injury and showed no response to various medications. The abdominal swelling due to the gas was eliminated shortly after starting on Chlorella, but her abdomen became swollen again with gas as soon as she stopped taking it.

CASE : 48 year old man with liver cirrhosis, abdominal gas distension and water retention. The severe abdominal gas and constipation were alleviated very soon after starting to take Chlorella.

Fasting With Chlorella

As a method of cleansing the body and rejuvenation, fasting has long been used. With its poison-absorbing qualities and its ability to stimulate the immune system, Chlorella can be used in a program where cleansing and immune stimulation is desired.

INFECTIONS

Anti-viral activity was first found in Chlorella by Shirota Minoru et. al. who found the activity within lipid fractions. It was found effective against viruses belonging to the entero and adeno groups in vitro[110].

Chlorella cells, high pressure processed cells and CGF were tested by another group of scientists for antibody stimulating ability and these doctors found that both T and B cells were activated. CGF was studied for its anti-viral action by giving mice 2 mg every other day for two weeks. Controls and experimental animals were then inoculated with a potent influenza virus directly into the brain. An increased survival rate and a prolonged survival time were seen in the mice pre-treated with CGF and an antivirus antibody[111].

Antiviral activity of Chlorella was also described by Fukada and Associates in 1968[112].

Graph of Virus Inactivation

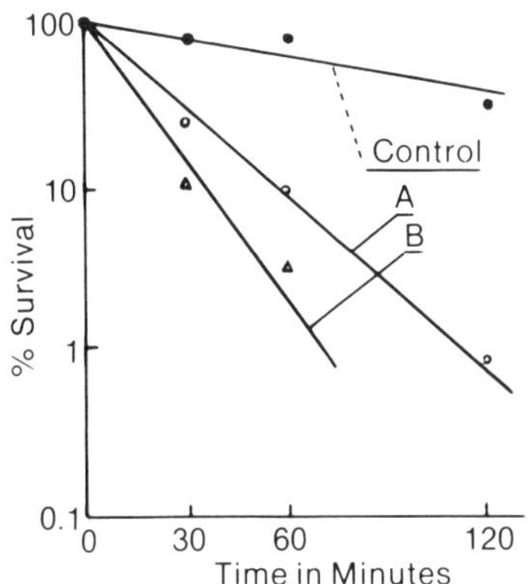

Fig.1   Inactivation of Sindbis virus by chlorella cell homogenate. The reaction mixture containing the algal component (A,60 µg/ml; B, 600 µg/ml; wet weight) and 10^7 Plaque-forming units per ml of Sindbis virus, was incubated at 30C; samples were withdrawn at indicated times, diluted with Eugle's medium, and assayed on a monolayer culture of chick embryo cells by the plaque technique

Chlorella induces higher levels of interferon and this interferon was demonstrated to be protective against vaccinia and influenza virus inoculations in mice[101]. As has been mentioned, interferon is a natural secretion of the body and is thought to be a physiological stimulator of macrophages[102, 103]. Injection of interferon enhances macrophages' oxidative metabolism and function[104].

Interferon has been shown to promote the killing of many kinds of intracellular bacteria: Toxoplasma gondi (**Toxoplasmosis**), Leishmania donovani (**Leishmaniasis**), Listeria monocytogenes (**Listeriosis**), Mycobacterium intracellulare (associated with **chronic lung disease** in Man), and Mycobacterium leprae (**leprosy**)[105].

In 1953, Chlorella was first used clinically to treat a medical condition. A Dr. Jorgensen in Venezuela gave a "soup" of vegetable and animal plankton to people suffering from an advanced form of **lepromatous leprosy.** They were of all ages: thirty-seven were from 8 to 20 years of age, twenty-six were from 20 to 40 years of age and seventeen were from 40 to 70 years of age. A large percent of this "soup" mixture consisted of Chlorella. No ill effects were seen; the patients were "glad to take it" and did so for one to three years. In a majority of the cases **there was a marked improvement in energy, in weight gain, and in general health,** all of which seemed to result from the plankton "soup."[150].

Chlorella has been used for **preventing pneumonia** in calves. Veterinarians studied a group of 453 calves between birth and one year of age and showed that those given Chlorella had a much lower incidence of bronchopneumonia[151].

Epstein — Barr virus is the cause of **infectious mononucleosis** and in certain cases may produce a chronic state of ill-health characterized by repeated infections, allergies and exhaustion. **Cytomegalovirus** may also produce a chronic illness that causes exhaustion and increased allergies. I have had the opportunity of giving Chlorella to a number of people with these disorders and **all have related that Chlorella helped them.** An antibiotic has been isolated from Chlorella and has been shown to be effective against some types of gram-positive and gram-negative bacteria in vitro (in the test tube). This antibiotic was named chlorellin and was found to result from photo-oxidation and splitting of long carbon chains of the unsaturated fatty acids[152-157].

This process has been used to advantage in treatment of sewage. During World War II it was observed at several armed services installations in California and Nevada that the effluent from open sewage settling pools heavily inoculated with Chlorella was bacteriologically safe for discharge into local streams. In fact, principally during the summer months when high light intensity favored the growth of the algae, the coliform count was frequently lower in effluent from pools heavily inoculated with Chlorella and not chlorinated than it was in effluent from pools chlorinated but without Chlorella[157, 159].

THE COMMON COLD

Probably the most extensive human test of Chlorella's antiviral activity was done back in 1966 on sailors of the Japanese navy while sailing from Tokyo to New Zealand during a 95 day period. The control group consisted of 513 crewmen and the experimental group consisted of 458 crewmen all between the ages of 19 and 30 who took 2 grams of Chlorella tablets per day. Six times during the voyage the men were checked for colds and the results tabulated. At the end of the trip those men taking Chlorella tablets had 571 colds while the control group had 903. Thus **26.5% less colds occurred in those taking two grams of Chlorella per day**[112].

A new, still experimental treatment for colds is intranasal administration of interferon. Perhaps one reason Chlorella helps prevent colds is because it produces high levels of interferon.

STOMACH AND DUODENUM

Stomach and duodenal problems are one of modern Man's most frequent health problems. Ulcers of these areas can be due to stress, alcohol ingestion, hot spices, coffee, etc. and the cause of the ulcer should be sought and eliminated. Chlorella has been studied in the treatment of stomach and duodenal ulcers with fairly good results.[159, 160].

SKIN PROBLEMS

Chlorella has been reported to be **useful in treatment of eczema, for prevention of recurrent cold sores, warts, allergic dermatitis, and acne.** The experiences of one of my research participants is illustrative.

ALLERGIES

Many cases of **allergic rhinitis, atopic dermatitis and asthma have been reported to have been helped** with long term administration of Chlorella. To my knowledge this has not been subjected to a scientific trial. Colds often weaken the body and act to initiate allergies and asthma. Chlorella's cold preventive actions may be one reason that allergies are helped. Another reason that it may help is the immune regulating action of the interferon that is induced by the taking of Chlorella. Chlorella is said to "normalize" body functions and it may well be due to the increased levels of interferon. Another reason may be the increased numbers and functions of the body's macrophages. By activating these cells the foreign proteins that cause allergies may be removed more quickly from the body which keeps these proteins from inducing an allergic response. Cleansing of the colon, improvement of liver detoxification mechanisms and Chlorella's content of vitamin A and Beta-carotene(repairs and strengthens mucous membranes) may all contribute to Chlorella's anti-allergy activity. That Chlorella has a beneficial effect on the liver and thereby allergies

— can be deduced from a study of long term prisoners. The men were divided into four groups: (1) control; (2) amino acid supplemented; (3) Chlorella supplemented (2 gram/day); (4) amino acids plus Chlorella. Before and after three months the prisoner's blood was analyzed for proteins and amino acids. Compared to pretreatment levels the albumen levels increased (a measure of the liver's ability to make protein) and gamma globulin (involved with allergies) levels fell. These changes are what would be expected if the liver is functioning better[161, 162].

This same type of finding (increased albumin, decreased globulin) was found by a Russian scientist in the treatment of young pigs with Chlorella[162].

Sindo et. al. reported the inhibitory effect of chlorophyll derivatives on complement activity and anaphylactic reactions[88].

More recently Sato et. al. have demonstrated that sodium copper chlorophyllin (a mixture of copper chelates of chlorophyll derivatives) **prevents lipid peroxidation** (oxidation of fatty membranes) and the breakdown of lysosomes in liver tissues[89].

Allergic reactions use a number of proteolytic enzymes in their production and it thus would seem likely that chlorophyll with its anti-proteolytic activity has anti-allergic properties. This may well explain in part the benefit that patients with food allergies and arthritis experience when they switch their diets to more green vegetables or take Chlorella.

Many of the people reported to have been helped by Chlorella **have taken it for at least 3 months before any noticeable improvement in their allergies occurred.** One lady wrote this letter to me about her experiences with Chlorella.

Sept. 11, 1986

Dear Doctor Steenblock;

I thank God everyday for making algae and for the Japanese people making it available. I started taking Chlorella in September 1976 and three years later Dr. Charles Jones, my eye doctor told me the crystalization on the lens of my eyes was completely gone. Three years ago I developed dry eyes which doctors have no cure for. I took 100 tablets a day for 2 days and I didn't have dry eyes anymore. I was afraid to stop taking 100 so I continued for 16 days then I gradually reduced to thirty. I was allergic to dogs, cats, dust and pollens. Dr. John Bookman said to me and I quote verbatum "You have to have this shot every week as long as you live." I had tests several years ago with Dr. Tanikin and I am not allergic to anything now! I had quit taking the shots in 1979. I had Menier's disease. Had to put drops under my tongue. The last time I had an attack I also had the flu which was in 1971. I had the flu several times every year since I was nine years old but I haven't had the flu since I started taking Chlorella (1976). I had two fatty tumors on my ribs which disappeared. I had arthritis so bad I had to wear a TENS unit to shock my nervous system and kill the pain. I also had to take codeine and butazolidine. I no longer have to use the TENS unit and am off of drugs. Two years ago I started to take the Chlorella extract and everyone tells me I look younger. No one believes I am 77. One of the sales girls at the convention center said to me "I don't believe you. I want to see your driver's license." Everyone guesses my age to be between 45 and 50 years old. I feel young and no one thinks of me as being old. I feel like I am sweet sixteen.

Treva Corbett
Los Angeles, CA

ARTHRITIS

Everyone knows that arthritis means painful joints but most don't know that arthritis can be due to over fifty different causes. The most common type is OSTEOARTHRITIS and has usually been thought to be due to wear and tear of the joints as they age. The lining of our joints is made up of a synovial membrane. This membrane is made up of one to three layers of lining cells — some of which have phagocytic activity (macrophages). The membrane allows easy diffusion of most materials from the blood into the joint spaces. The capillaries and small veins contained within the membrane have small openings which allow material to pass through easily into the joint space from the blood[171]. When poisons (bacteria, viruses, mycoplasma, chemicals) from the intestine enter into the blood they may settle into these joint spaces and cause an immune response. People with RHEUMATOID ARTHRITIS have cells (macrophages and lymphocytes) that are especially sensitive to viruses and other poisons and are stimulated to become more active by these. Once a poison has "turned on" the joint's macrophages and lymphocytes — arthritis begins which leads to damage to the joint surfaces and scarring and calcification are produced. The subsequent changes in the joint cause an increased hardness of the tissues that surround the delicate blood vessels which supply the joint surfaces. Normally the blood vessels contained within the joint have some ability to expand and contract. When this ability is lost, blood particles such as red cell and platelet-leucocyte clumps are unable to pass through the joint's blood vessels as well. These blood vessels then become blocked for short periods of time causing further damage to the joint by creating temporary episodes of lack of oxygen to the area. Food allergy induced arthritis is a good example of this process. The arthritis that occurs with inflammatory bowel disease is another example of this type of reaction. In this case there can be circulating aggregates of immune complexes, bacteria, red cell and monocyte aggregates and serum antibodies that develop against certain bacteria, fungi and other organisms

29

that are able to enter the blood through the damaged intestinal wall. These antibodies cross-react against the joint tissues and damage the joints. This process may occur in many types of arthritis and correction of intestinal problems often helps correct the arthritis. Chlorella strengthens the intestine and helps clear away toxic substances. This allows the body to protect itself from the entrance of bacteria, fungi, and poisons into the body through the bowel wall thereby preventing their damaging effects on the joints. Chlorella strengthens the function of the liver and increases its ability to remove bacteria, fungi and toxins from the blood. The liver then excretes these poisons into the intestine where they may be reabsorbed in the lower part of the small bowel. The less fiber a person has in their diet, the slower and weaker the bowel will become. When this occurs the more absorption of poisons can and does happen since the bowel wall gradually grows thinner. These poisonous substances have been implicated in cancer and most probably are involved with arthritis as well. Chlorella cell walls may absorb these substances but no studies have been done to prove this.

RHEUMATOID ARTHRITIS is the crippling type of arthritis and is due to an overactivity of one part of the immune system and a dysregulation of another part. In rheumatoid arthritis synovial membrane tissue proliferates and destroys joint cartilages, bones and tendons. **Collagenase** is a major agent in the destruction of these tissues which **is inhibited by Beta-carotene.** Chlorella has a high vitamin A and Beta-carotene level and therefore may help suppress the inflammation due to rheumatoid arthritis[107].

As mentioned previously, Chlorella stimulates interferon production. Recent studies have demonstrated that interferon-gamma has great therapeutic effects on rheumatoid arthritis. In the first study reported in the literature which used interferon for treatment of rheumatoid arthritis, 80 patients were given interferon and 58 showed considerable improvement with pain decreasing rapidly and often disappearing[109]. Even in largely immobile patients the ability to walk and general mobility were nearly fully restored!

This effect of interferon demonstrates indirectly a property of Chlorella which is that it acts to normalize the body. On the one hand it has been shown to stimulate macrophages and these cells are responsible for much of the damage in rheumatoid arthritis. On the other hand interferon formation is induced by Chlorella and interferon improves arthritis rather than making it worse. This is because interferon is a cell regulator as well as being a stimulator. Rheumatoid arthritis is a disease of immune dysregulation which apparently can be corrected by interferon (by injection) and possibly by its increased production when Chlorella is taken orally.

Arthritis is an acid condition of the joints and Chlorella has been claimed to be effective in neutralizing acidic constitutions. By raising the joint's pH, the overactivity of the white cells contained within the painful joints decreases and pain is reduced.

Many people claim that Chlorella has helped their gout. Gout is caused by elevated levels of uric acid which precipitates in the joints and tissues. In one study done at the department of Nutritional Sciences at the University of California at Berkeley, 50 grams of Chlorella per day was shown to increase the urinary excretion of uric acid but was associated with elevated blood levels of uric acid[109]. Further work needs to be done on establishing Chlorella as a treatment for gout and other forms of arthritis. The finding that Chlorella may elevate the blood level of uric acid should be considered if a person has had gout since high blood levels of uric acid are associated with attacks of gout.

In patients with arthritis and especially those with chronic bowel problems the use of Chlorella should be in small doses initially and increased very gradually and only as the person tolerates it without pain, diarrhea or cramps.

ATHEROSCLEROSIS AND HARDENING OF THE ARTERIES

Atherosclerosis is the process of gradual narrowing and hardening of the arteries of the body secondary to the wear and tear on the blood vessels by the irritative components (chiefly cholesterol) of the blood. By keeping the blood cholesterol at a level below 160 mg% a person is generally able to avoid a heart attack or other degenerative artery disease(s). By keeping the blood cholesterol below 160 mg% for a two year period, actual reversal of atherosclerosis has been known to occur. Because of these facts modern day physicians are constantly seeking out ways to lower the blood level of cholesterol to as low a level as possible. Many drug companies have spent millions of dollars in hopes of finding the one drug that would lower cholesterol levels safely for long periods of time. Certain studies with Chlorella would indicate that it may play a role in improving the function of aging blood vessels that are affected by atherosclerosis.

Hematoporphyrin derivative (described in the section on Cancer — page 9) is similar to some of the chemicals contained in Chlorella. It was recently shown that when hematoporphyrin is injected into animals with atherosclerosis, it is selectively picked up by and absorbed into atherosclerotic plaques. Atherosclerotic plaques contain smooth muscle cells which cease to grow when exposed to this substance[105] bringing up the possibility that **Chlorella may prevent atherosclerosis by inhibition of arterial smooth muscle cell proliferation.**

CHOLESTEROL

Injected chlorophyll has a stimulatory effect on apo-lipoprotein synthesis in rat liver and on excretion of the free form of cholesterol into the bile in vitro[59]. This may explain in part the mechanism by which Chlorella lowers the levels of blood cholesterol since increased excretion of cholesterol into the small intestine where it passes out with the stool will lower the blood cholesterol concentration.

The first report that I am aware of that used Chlorella to lower blood cholesterol levels was reported in 1969. In this study, rabbits were given papain, vitamin B-5, and Chlorella polysaccharide orally and it was shown that Chlorella lowered cholesterol levels as well as the neutral fat content of the rabbits' blood[134].

When Chlorella is given to rats, no significant change is seen as long as the cholesterol level is already low but as the cholesterol level rises, Chlorella causes a drop in the serum cholesterol (216 mg% dropped to 165 mg% on 8% dietary Chlorella)[135].

Dr. T. Sakuno reported also that Chlorella powder "significantly inhibited" the rise in rat's serum cholesterol when fed a high cholesterol diet[136].

Rats placed on a high cholesterol diet were fed 10% Chlorella and subsequently showed a significant inhibition in the rate of rise of serum cholesterol[137].

Two other studies showed a drop in the cholesterol in serum and livers with the administration of Chlorella[138, 139].

HIGH BLOOD PRESSURE

"It is known by experience that Chlorella is effective against hypertension and fatigue"[140].

Many case reports have been documented of people claiming that their blood pressure problems disappeared after taking Chlorella for durations of 3 months to 1 year.

Chlorella extracts have been shown to lower the blood pressure of Wistar rats after intravenous administration. The blood pressure showed an average fall of 63 mmHg one hour after intravenous administration and a 32 mmHg average fall in rats with normal blood pressure[142].

In another study spontaneously hypertensive rats were fed Chlorella for 7 weeks and **the animals' blood pressure became "significantly lower" than the controls** and remained 20 - 30 mmHg lower throughout the remaining 23 weeks of the experiment. Renin levels were measured in some of the rats after 25 weeks and found to be significantly lower in the Chlorella treated group (16 + or − 5.45 VS 40.77 + or − 8.9 ng/hr/ml). Renin is a kidney-produced hormone that elevates blood pressure[143].

In a commissioned study of Chlorella on the human pulse wave velocity, Dr. Shimizu and Associates of Kanazawa Medical University studied 10 people over a 2 month period. These individuals received about 1/2 ounce (14cc) of Chlorella Extract and 0.25 gram (1¼ tablet) of autoclaved Chlorella tablets per day and had their pulse wave velocity (an index of blood vessel elasticity) measured before and after treatment. The results showed a tendency for the pulse wave velocity to become slower, suggesting improved vascular elasticity, in five of the ten persons. However the dose used in this study was very small and the duration of the experiment may have been too short for the full effect of the Chlorella to be seen in the structure and function of blood vessels[144].

In 1932 Dr. Burgi published his book entitled "Chlorophyll as a Pharmaceutical", in which he related his treatment with chlorophyll of 112 patients afflicted with hypertension associated with arteriosclerosis (hardening of the arteries) and showed that in most cases the blood pressure returned to normal[145].

In 1931 Angelo studied 50 people by dividing them into 3 groups. Group I had normal blood pressure (10 people), Group II had high blood pressure (17 people) — these were hospitalized due to complications of their high blood pressure, and Group III was 23 people suffering from high blood pressure, arteriosclerosis and nephrosclerosis but living at home. These people were given water-soluble chlorophyll which was given in doses of 0.18 to 1.5 grams per day to the total exclusion of all other pharmacological treatments. Angelo concluded: **"By studying the action of chlorophyllin on the blood pressure of normal and hypertensive subjects, the following observations were made: 1. Chlorophyll is well tolerated and does not produce toxic symptoms even when administered in large doses and for several months. 2. It does not have any special effect upon the blood pressure of normal subjects. 3. In hypertensive patients it causes a steadily progressive fall in the arterial pressure and the improved condition of the blood pressure is accompanied by a feeling of well-being"**[146].

In 1932 Zickgraf reported "Of the 18 cases treated by myself, only two cases of arteriosclerosis of the brain were not influenced in respect to blood pressure. As for the effect of chlorophyllin in the case of arteriosclerosis, many publications exist. Lowering of the blood pressure by widening of the arterial vessels, longlasting effect of chlorophyll, strengthening of the heart strength, besides the general tonic properties of the chlorophyllin, has been observed."[147].

Chlorella has been used to promote the regeneration of wound tissue such as found in diabetic ulcerations and ulcerations due to atherosclerosis and venous insufficiency[148, 149].

TOXICOLOGY

Vitamin A toxicity is theoretically possible from Chlorella but one would have to consume 300 tablets or 20 packets (60 grams) /day to get the 48,000 units of vitamin A per day needed to be taken for many months to get poisoned from the vitamin A.

The FDA's select committee reviewing the status of brown and red algae stated "in the Far East, seaweeds have been an accepted food for humans for centuries, constituting up to 25% of the diet."[34].

The Huntington Research Centre in England conducted studies on the acute oral toxicity to rats of green Chlorella and yellow Chlorella powders. Rats were fed varying doses of Chlorella powders and after 14 days of observation, the rats were killed and examined macroscopically. The study concluded that the acute median lethal oral dose (LD50) was in excess of 16 gram/kilogram bodyweight (2). This would equal over 1 kilogram (2.2 pounds) for the average 70 kilogram (150 pound) person that would have to be taken at one time and even at that dose the rats showed no evidence of toxicity. In other words **no toxicity was found at the highest doses of Chlorella tested**(165).

Kyowa Hakko fed 50 five week old rats graduated amounts of pulverized Chlorella for 90 days. Blood and urine tests were done on 20 which were all normal and the remaining animals were autopsied and found to have normal organs(166).

Another study used young rats that were fed pulverized Chlorella in graduated amounts instead of protein. No differences were found between experimental and control animals in terms of growth or organs except that the livers of the control group were slightly lighter. No evidence of toxicity was found(167).

Experiments have been done in which human volunteers have eaten a diet containing no protein other than that from Chlorella for 20 days with no ill effects seen(168). In research here at the Aging Research Institute volunteers were studied for 6 weeks while taking 15 tablets per day. Multiple blood chemistries and blood counts were done before, during and after taking the tablets and no statistically significant change in any laboratory test was found. On the contrary most of the volunteers reported positive health benefits. These were some of the volunteers comments:

"helped the vaginal yeast problem."

"My hands and feet were cold before, but the tablets helped them get warmer"

"I have decreased cravings for sweets"

"My hair is no longer dry and has been improved"

"reduced food cravings, heart burn, bad breath"

"reduced headaches"

"feel more alert, better memory, improved energy"

"reduced coating on my tongue"

"reduced daytime sleepiness"

"cleaned up the tarter on my teeth" (patient chewed the tablets).

Although not designed to demonstrate the safety of Chlorella, a study was done that used Chlorella as a dietary supplement for pregnant women suffering from iron deficiency anemia. In those women given Chlorella extract, the serum iron increased more than in the others(169). If nothing else this study illustrates how safe the scientists thought Chlorella to be, since giving any substance to a pregnant woman is inviting a lawsuit if the child is born with any type of physical problem. No report of any detrimental effects on the offspring was mentioned.

As mentioned earlier, there have been cases of photosensitivity (skin reactions triggered by light) reported after ingestion of chlorella products from one company that had problems with production some 10 years ago[175, 182, 183]. As mentioned in the section on cancer, pheophorbide is the chlorophyll breakdown product that is responsible for the light sensitivity. The Japanese government has established regulations on the permissible amount of pheophorbide in Chlorella products. This is 160 mg/100 gram of product. One product available here in the U.S.A. contains only between 40 - 50 mg/100 gram. It is thus theoretically possible that if a person takes huge quantities of Chlorella, and is lacking vitamins C, E, beta-carotene, vitamin B-5 and coenzyme Q and is exposed to sunlight — a photosensitivity reaction could occur. In rats treated with known quantities of pheophorbide the LD 50 was 45.5 mg/100 gram body weight and the MLD was 12 mg/100 gram body weight[181]. Using the 12 mg/100 gram body weight figure, the average 70 kilogram person would have to consume 8,400 mg pheophorbide or 16,800 grams of Chlorella. (70 kilos/100 = 700. 700 × 12 = 8,400 mg. At 50 mg pheophorbide/100 gram Chlorella 8,400/40 = 168 × 100 = 16,800 gram of Chlorella). To get enough pheophorbide to cause a problem would thus be almost impossible.

Chlorella has been marketed in Japan since 1963, in Hawaii and Taiwan since 1970, and in Canada since 1978. From 1974 to 1977 the total world consumption was over 500 metric tons yearly. It is currently the most popular health food supplement in Japan and the yearly production has risen to over 1000 tons, all of which is used for human consumption. The number of people taking Chlorella daily on the average is approximately 3 million people[166].

POSSIBLE REACTIONS

Chlorella may be taken alone or with medications.

The following initial reactions may be noted:

1. Intestinal gases may be released due to rejuvenation of the peristaltic action of the intestines. This will cease as the intestines become cleansed.

2. Irregularity of bowel movement, nausea or slight fever may be noticed in a small number of people. This symptom usually disappears within 2 - 3 days, but very occasionally may take up to ten days to disappear. These reactions are most prevalent in those persons who need Chlorella the most since these people have the weakest bowels and have greater toxicity to neutralize.

3. Allergy sufferers sometimes break out in pimples, rashes, boils or eczema, in some cases accompanied by itching. This means that the drive to regain its internal balance is being accelerated and the body is actively working to expel toxins.

4. Bowel movements may become greenish in color. This simply means that excess chlorophyll is being expelled but doesn't mean you are taking too much Chlorella.

The above reactions should not be taken as side-effects but as favorable reactions which appear as part of the body's adjustment process while taking Chlorella. These indications show that the Chlorella is working. **(AUTHOR'S NOTE: The possibility of a true allergic reaction is always possible and therefore if you are having severe reactions it may be the prudent course to stop the Chlorella for a few days before attempting to take it again. If only one to three tablets gives you problems then you may well be allergic and consultation with your doctor should be done to determine if you should continue taking Chlorella).**

RECOMMENDED DOSAGE

Although you may take as many Chlorella tablets per day as you like, it is always better to start with 1 to 5 with a meal three times per day and gradually increase the amount.

1. For general prevention and maintenance: approximately 15 to 20 tablets per day.

2. When actual symptoms become apparent indicating the necessity of Chlorella treatment: take 25 to 30 tablets per day in divided doses. If stomach distress is present, use the Chlorella granules, mix with pure water and use a blender to mix all of the particles into solution.

3. For optimizing Chlorella's detoxification properties, take either the tablets or granules orally on an empty stomach.

Chlorella can be taken at one time or divided into smaller amounts several times per day before meals or on an empty stomach.

translated from: Japan Chlorella Research Center, Kyoto, Japan.

SUMMARY

Chlorella — so much potential in such a small structure! It seems this tiny algae is beginning to "bear fruit" for the human race after surviving on earth for so many millions of years.

The author has reviewed most of the known scientific papers on the medical and health uses of Chlorella. We have seen that it has anti-viral activity protecting against the common cold. Furthermore it has been shown to increase the growth rate of children and to make them healthier and stronger. It is beginning to be used clinically for detoxifying people who have been exposed to heavy metals, insecticides, pesticides and hydrocarbons with good results. It has been used to treat infected and draining skin ulcers with success. Pancreatitis may be helped and alcoholic hangovers are claimed by one scientist to be prevented by the taking of Chlorella. It is useful in eliminating body odors and restores the normal functions of the bowels. The immune system has been shown to be stimulated by its use and interferon levels increase. With its high vitamin A and beta-carotene content it gives hope for effectiveness in prevention of cancer. Research is showing promise of its usefulness in preventing the spread of cancer as well as helping the body fight this dread disease once it has become established. Chlorella has been shown to lower the chlolesterol levels and in preliminary studies has been shown to soften blood vessels that had become hardened with age and deterioration.

With the continuous exposure of our foods to hydrocarbon insecticides and herbicides it seems that Chlorella should find its way onto everyone's dining room table. Chlorella should be taken daily to remove these harmful substances from our bodies before they cause cancer and other degenerative diseases. Chlorella has great potential in aiding the treatment of cancer, AIDS, Epstein Barr virus and other chronic viral illnesses and is especially helpful in preventing the degenerative diseases associated with the aging processes.

MESSAGE TO READERS

The Aging Research Institute is a non-profit corporation that studies the diseases associated with the aging process and how to reverse them by use of natural substances. Your contribution will help our goals of eliminating cancer and atherosclerosis and ultimately to halt or even to reverse the process of aging itself. Any contribution no matter how small, is greatly appreciated. Thank you for buying this book and supporting our research efforts. Send your contribution or correspondance to:

The Aging Research Institute
22821 Lake Forest Drive, Suite 114
El Toro, California 92630

Your personal experiences with Chlorella would be most appreciated and should be typed in detail and sent to the above address with a signed statement that includes your printed name and that states you give express permission to use your letter. Please consult your own physician to determine if this substance will be appropriate for you. NO CORRESPONDENCE WITHOUT A SELF-ADDRESSED, STAMPED ENVELOPE WILL BE ANSWERED.

David A. Steenblock, M.Sc., D.O.

INDEX

BIBLIOGRAPHY

(1) Bewicke, Dhyana and Beverly A. Potter. Chlorella, The Emerald Food. Ronin Publishing Inc. Berkeley, California 1984.

(2) Beyerinck, M.W. Culturversuche mit Zoochlorellen, Lichenengonidien, und anderen niederen Algen. Bot.Z. 48, 725, and following articles. 1890.

(3) Schopf, J. William. Precambrian Micro-organism and evolutionary Events Prior to the Origin of Vascular Plants. Biol.Rev. (1970), 45, pp. 319-352.

(4) Kirk, R.E., and D.F. Othmer. Encyclopedia of Chemical Technology. Vol. 3, p.879, The Interscience Encyclopedia, Inc., New York, 1949 (2nd ed.).

(5) Burgi, E. Cor.-bl. f.Schweiz. Aerzte, 46,449-66, 1916.

(6) Burgi, E.: Das Chlorophyll als Wachstumsstoff, Klin. Wchnschr. 9:789, (April 26), 1930.

(7) Burgi, E.:Das Chlorophyll als Pharmakon, Thieme, Leipzig, 1932.

(8) Gordonoff, T. and Ludwig, F.: Ueber den Einfluss der Vitamine auf das Wachstum von Gewebe und von Impfgeschwulsten. Ztschr. f. Vitaminforsch. 4:213-223 (July) 1935.

(9) Burgi, E.: Wirkungen von Pflanzenfarbstoffen auf die verletzte Haut, Schweiz. med. Wchnschr. 67:1173-1176 (Dec. 11) 1937.; Ueber die Wirkung von Chlorophyll auf die Wundheilung, ibid. 68:483-485 (April 30) 1938.

(10) Gahan, E., P.R. Kline and T.H. Finkle. Chlorophyll in the Treatment of Ulcers. Arch Dermatol. Syphilol. 49, 849-851, 1943.

(11) Hughes, J.H. and A.L. Latner. Chlorophyll and Haemoglobin regeneration after haemorrhage. J. Physiol. 86, 388-395, 1936.

(12) Noack, K. and Kiessling. The origin of chlorophyll and its relation to the blood pigment. Z. physiol. Chem. 182: 13, 1929. Cited in Chem abstracts 23:3249.

(13) Kephart, J.C.: Chlorophyll Derivatives-Their Chemistry, Commercial Preparation and Uses. Econ. Bot. 9, 3-38, 1955.

(14) Gruskin, B.: Chlorophyll: Its Therapeutic Place in Acute and Suppurative Disease, Am.J.Surg. 49,49-55 (July), 1940.

(15) Bowers, W.F., Military Surgeon, 90, 140-52, 1942.

(16) Goldberg, S.L.: The Use of Water Soluble Chlorophyll in Oral Sepsis. Am.J.Surg. 62, 117-23, 1943.

(17) Smith, L.W. and A.E. Livingston. Chlorophyll. An experimental study of its water soluble derivatives in wound healing. Amer. J. Surg. New Series vol. LXII, No. 3, 358-369, Dec. 1943.

(18) Smith, W.L., and Livingston A.: Wound Healing. Am. J.Surg., 67,30-39, 1945.

(19) Yamagishi, Y., Hasuda, S., Y.Mito. V. Experience in taking Chlorella for healing the less curable wound. In: Huang, C.J.:Application of Chlorella on Medicine and Food. Technical Bulletin, March 1970. Taiwan Chlorella Manufacture Co., Ltd., Taipei, Taiwan.

(20) Usenko, G.V. [Comparative evaluation of various methods for the overall treatment of inflammatory diseases of the internal female genitalia using chlorophyllypt.] Pediatr. Akush.Ginekol. (5), 57-9, Sep-Oct. 1974. Ukranian.

(21) Honek, L. et. al. [The use of a fresh water weed Chlorella vulgaris for the treatment of the cervix after Kryo-surgical interventions.] Cesk. Gynekol. 43(4):271-3, May 1978.

(22) Holmes, G.W. and H.P. Mueller. Treatment of Post-Irradiation Erythema with Chlorophyll Ointment. Am. J. Roentgenol. Rad. Ther. 50,210-213, 1943.

(23) Smith, L.W.: Chlorophyll: An Experimental Study of its Water Soluble Derivatives. Am.J. Med. Sci.207, 647-654, 1944.

(24) Smith, L.W.: Chlorophyll: An Experimental Study of its Water-Soluble Derivatives. IV. The Effect of Water-Soluble Chlorophyll Derivatives and Other Agents Upon the Growth of Fibroblasts in Tissue Culture. J.Lab. Clin. Med. 28,241-246, 1944.

(25) Brocklehurst, J.C.: An assessment of chlorophyll as a deodorant. Brit. Med. J. 1:541, 1953.

(26) Golden, T. and Burke, J.F.:Effective management of offensive odors. Gastroenterology, 31:no.3, 1956.

(27) Montgomery, R.M. and Nachtigall, H.B.: Oral administration of chlorophyll fractions for body deodorization. Postgrad. Med., 8:501, 1950.

(28) Westcott, F.H.:Oral chlorophyll fractions for body and breath deodorization. N.Y.J. Med., 50:638, 1950.

(29) Morrison, J.E.:Oral tablets help control ward odors, study shows. J.Amer. Hosp. Assoc. July 16, 1959.

(30) Young, R.W. and J. S. Beregi. Use of Chlorophyllin in the Care of Geriatric Patients. J. Am. Geriatrics Soc. XXVIII, No.1, p.46-47, 1980.

(31) Weingarten, M. and Payson, B.: Deodorization of colostomies with chlorophyllin. Rev. Gastroenterol. 18:602, 1951.

(32) Young, R.W. and J.S. Beregi. Use of Chlorophyllin in the Care of Geriatric patients. J. Am. Geriatrics Soc. XXVIII, No.1, p.46-47, 1980.

(33) Killian, J.A. and Panzrella, F.P. Comparative studies of samples of perspiration collected from clean and unclean skins of human subjects. Toilet Goods Assoc. Proc., Sci. Sect. 7:3, 1947.

(34) 1978 Food Drug Cosmetic Law Reports (CCH), paragraph 45, 613, at 47, 597.

(35) Nedwin, G.E., Svedersky, L.P., Bringman, T.S., Palladino, M.A.Jr., Goeddel, D.V.: Effect of interleukin 2, interferon-gamma, and mitogens on the production of tumor necrosis factors alpha and beta. J.Immunol. 1985; 135:2492-7

(36) Sugarman, B.J., Aggarwal, B.B., Hass, P.E., Figari, I.S., Palladino, M.A. Jr., Shepard, H.M.:Recombinant human tumor necrosis factor-alpha:effects on proliferation of normal and transformed cells in vitro. Science 1985;230:943-5.

(37) Strum, W.B. and H.M.Spiro. Chronic Pancreatitis. Annals of Internal Medicine 74:254-277, 1971.

(38) Yoshida, A. et. al. Gastroenterology Jpn 1980;15(1):49-61.

(39) Manabe, T. et. al. Ann. Surg. 190 (1):13-7, Jul 1979.

(40) Mann,S.K. et. al. Arch. Pathol. Lab. Med. 103(2):79-81, Feb. 1979

(41) Wiznitzer, T. et. al. Am. J. Dig. Dis 21 (6):459-64, Jun 1976.

(42) Hagiwara, Y. Green Barley Essence. Keats Publishing, Inc., New Canaan, Connecticut, 1985. pp128-130.

(43) Oda, T., Yokono, O., Yoshida, A., Miyake, K., Iino, S.:On the successful treatment of pancreatitis with chlorophyll-a and inhibiting effects of its derivatives on trypsin and other protease activities in vitro. Gastroenterol. Jpn 6:49, 1971.

(44) Yoshida, A., O. Yokono and T. Oda. The Effects of Intravenously Administered Chlorophyll-a on Naturally Occurring Serum Protease Inhibitors in Rabbits. Gastroent. Jpn. 15(1),41-48, 1980.

(45) Oda, T., Yokono, O., Yoshida, A. et. al:The effect of chlorophyll-a in the treatment of pancreatitis. Nihon Iji Shimpo 2438, 1969 (inJpn)

(46) Oda, T., Yokono, O., Yoshida, A. Miyake, K., Iino, S.:On the successful treatment of pancreatitis with chlorophyll-a and inhibitory effects of its derivatives on trypsin and other protease activities in vitro. Gastroenterol Jpn. 6:49, 1971.

(47) Mann, N.S.: Experimental acute hemorrhagic pancreatitis; Effect of chlorophyll-a. Gastroenterology 74:1061, 1978.

(48) Orda, R. Wiznitzer, T.Bawnik, J.B. Bubis, J.J.:Effect of chlorophyll-a in experimental acute pancreatitis. Israel J. Med. Sci. 10:630, 1974.

(49) Wiznitzer, T., Orda, R., Bawnik, J.B., Bubis, J.J.: Acute necrotizing pancreatitis in the guinea pig-Effect of chlorophyll-a on survival times. Digestive Diseases 21:459, 1976.

(50) Mandai, H., Tsurumi, T., Harada, H., et. al:Remarkable effect of chlorophyll-a on a case with chronic relapsing pancreatitis. Gendai Iryo 6:1017, 1974(inJpn).

(51) Okano, T., Narasaki, M., Ueda, H.: Some experiences of chlorophyll-a in the treatment of pancreatitis. Gendai Iryo 6:1025, 1974 (in Jpn).

(52) Kawauchi, H., Kamijima, A., Namiki, M.:Therapeutic effects of chlorophyll-a against pancreatitis. Diagnosis and Treatment 59:156, 1971 (in Jpn).

(53) Kubo, K., Harada, Y., Abiko, H., Iwatake, T., Tamechika, N:Experience of chlorophyll-a in the treatment of pancreatitis. Diagnosis and Treatment 66:1100, 1978.

(54) Miyatake, R., Yoshida, S., Masujima, T., Fujita, S., Nakamura, M., Shimizu, T.:Clinical experience of chlorophyll-a in the treatment of chronic pancreatitis. Diagnosis and Treatment 66:1250, 1978 (in Jpn).

(55) Yasuoka, H., Takada, M., Matsuoka, Y., Ozaki, N.:Case report of severe acute pancreatitis treated with chlorophyll-a. Gendai Iryo 10: 1387, 1978 (in Jpn).

(56) Yamamoto, T., Koseki, Y., Miyagawa, F.:Inhibitory effects of chlorophyll-a on the pancrease alpha-amylase activity. Medicine and Biology 91:271, 1975 (in Jpn).

(57) Yamamoto, T., Miyagawa, F.:Inhibitory effect of chlorophyll-a on a pancreatic lipase. Medicine and Biology 92:11, 1976 (in Jpn).

(58) Imrie, C.W., Benjamin, I.S., Feruson, J.C., et. al:A Single-centre Double-blind Trail of Trasylol Therapy in Primary Acute Pancreatitis. Br. J.Surg., 65:337, 1978.

(59) Yoshida, A., O. Yokono and T. Oda. Therapeutic effect of chlorophyll-A in the treatment of patients with chronic pancreatitis. Gastroenterologia Japonica 15 (1), 49-61, 1980.

(60) McCutcheon, A.D. and Race, D.:Experimental Pancreatitis:Use of a New Antiproteolytic Substance, Trasylol. Ann. Surg., 158:233, 1963.

(61) Taoka, Y.:Studies on absorption, excretion and distribution in the body of exogenous chlorophyll-a. Igakunoayumi (Advances in Medicine) 91-241, 1974.

(62) Konno, K., Takeda, M.:Experimental report of absorption, excretion and metabolism of chlorophyll-a. The Clinical Report 8:2397, 1974.

(63) Miyagawa, F., Noguchi, T.:Effect of pantethine and chlorophyll-a to the fatty liver induced with the carbon tetrachloride. The St. Marianna Medical Journal 6:379, 1978.

(64) Nencki, M., and L. Marchlewski. Ber. deutsch. chem. Ges. 34:1687. 1901.

(65) Wilstatter, R. and M. Fischer. Chlorophyll XXIII. Die Stamm-substanzen der Phylline and Porphyrine. Ann. d. Chem. 400:182, 1913.

(66) Euler, H.v. and A. Stern. Die Chemie des Pyrrols. Vol. II, 2. Akadem. Verlagsgesellschaft, Leipzig. 1940.

(67) Granick, S. Protoporphyin 9 as a precursor of chlorophyll. J. Biol. Chem. 172:717, 1948.

(68) Granick, S. Magnesium protoporphyrin as a precursor of chlorophyll. J. Biol. Chem. 172:717. 1948.

(69) Granick, S. and H. Gilder. The Structure, function and inhibitory action of porphyrins. Science 101:504, 1945.

(70) Granick, S. and H. Gilder. The porphyrin requirements of Haemophilus influenzae and some functions of the vinyl and propionic acid side chains of heme. J.Gen. Physiol. 30:1, 1946.

(71) Hagino et. al. Effect of chlorella on fecal and urinary cadmium excretion in "Itai-itai," Jpn J Hyg 30(1), 77, April 1975.

(72) Pore, R.S.: Detoxification of chlordecone poisoned rats with chlorella and chlorella derived sporopollenin. Drug-Chem-Toxicol. 1984, 7(1), 57-71.

(73) Sassen, A., A. Van Eyden-Emons, A. Lamers and F. Wanka (1970):Cytobiologie 1: 273-382.

(74) Atkinson, A.W., D.E.S. Gunning and P.C.L. John (1972):Planta 107:1-32.

(75) Northcote, D.H., K.J. Goulding, and R.W. Horne.: The Chemical Composition and Structure of the Cell Wall of Chlorella pyrenoidosa. Biochem. J. 70:391-397, 1958.

(76) Horikoshi, T., A. Nakajima and T. Sakaguchi : Uptake of Uranium by Various Cell Fractions of Chlorella regularis. Radioisotopes 28(8), 485-487, Aug. 1979.

(77) Fink, H. Herold, E.: The quality of the protein of unicellular green algae and their effect in preventing liver necrosis. Zeitschr. Physiol Chem 305; 182-191, 1956.

(78) Want, L.F., J.-K. Lin and Y.-C. Tung. Effect of chlorella on the levels of glycogen, triglyceride and cholesterol in ethionine treated rats. J. Formosan Medical Association 79 (1), 1-10, 1980.

(79) Vermeil, C., O. Morin and L. LeBodic. (Anti-tumoral vaccination by peritoneal injection of micro-vegetable (yeasts and unicellular algae). Conceptual Error or Reality?) Archives Medicales de L'Oest Tome 14; no.10, pp.423-426.

(80) Takechi, Yoshiro. "Chlorella—Its Basis and Application. Publ. by Gakushu Kenku-Sha, Tokyo, Japan Nov. 30, 1971.

(81) Smith, L.W., and Spaulding, E.H.: The bacteriocidal and bacteriostatic action of water soluble chlorophyll a upon standard pathogenic cultures. Am. Jour. Med. Sci. 207:647, 1944.

(82) Rei Bunso, Body Revolution. Available from Hokkaido Medicinal Plant Research Institute.

(83) Silov, V.M., Liz'ko, N.N., Fofanov, V.I. and Kljuskina, N.S. (Effect of a diet containing destroyed alga cells on the microflora of the intestine) Kosmic. Biol. Med., 1969, No. 6, 54-57. Russian with English summary.

(84) Saito, Tatsumi, Saito Taku, and Oka, T.:Clinical Applications of chlorella pills. Medical examinations and new drugs. 3.3 pp.61-64, 1966, Japan.

(85) Nathan, DF, Murray, HW, Wiebe, ME, Rubin, BY. Identification of interferon-gamma as the lymphokine that activates human macrophage oxidative metabolism and antimicrobial activity. J.Exp. Med. 1983; 158:670-89.

(86) Nathan, CF. et. al. Local and Systemic Effects of Intradermal Recombinant Interferon-Gamma in Patients with Lepromatous Leprosy. N.Engl. J. Med. 1986; 315-15.

(87) Murray, H.W., Spitalny, G.L., Nathan, C.F.:Activation of mouse peritoneal macrophages in vitro and in vivo by interferon-gamma. J. Immunol. 1985; 134:1619-22.

(88) Sindo T., Haga K., Fuji G., Nishioka K.:Studies on the inhibition of chlorophyllin derivatives against complement activities and anaphylactic reaction. Jpn. J. Exp. Med. 36:489-498, 1966.

(89) Sato, M., K. Konagai, T. Kuwana, R. Kimura and T. Murata. Effect of Sodium Copper Chlorophyllin on Lipid Peroxidation. VII. Effect of Its Administration on the Stability of Rat Liver Lysosomes. Chem. Pharm. Bull. 32(7)2855-2858, 1984.

(90) Kojima, M., T. Kasajima, Y. Imai, S. Kobayashi, M. Dobashi and T. Uemura. A New Chlorella Polysaccharide and Its Accelerating Effect on the Phagocytic Activity of the Reticuloendothelial System. Recent Adv. R.E.S. Res. 13:11, 1973.

(91) Neveu, P.J., O. Morin, M. Miegeville, B.P. LeMevel and C. Vermeil: Modulation of antibody synthesis by an anti-tumour alga. Experientia 34/12, p. 1644-1645, 1978.

(92) Yamaguchi, N., S. Shimizu, T. Murayama, T. Saito, R.F. Wang, and Y.C. Tong: Immunomodulation by single cellular algae (Chlorella pyrenoidosa) and anti-tumor activities for tumor-bearing mice. Third International Congress of Developmental and Comparative Immunology, Reims, France, July 7-13, 1985.

(93) Kobayashi, S.: J. Japenese Reticuloendothelial Society 10, 83, 1971.

(94) Kobayashi, S.:Agricultural Chemistry 46, 373, 1972.

(95) Kobayashi, S.: Influence of chlorella extract on reticuloendothelial phagocytosis of rats. Health and Industry Newsletter, March 25, 1978. Agricultural Chemical Convention, Japan.

(96) Nakamura, M. et. al.: Promotion of reticuloendothelial function by chlorella components. Health and Industry Newsletter, March 25, 1978. Agricultural Chemical Convention.

(97) Hamada, M., M. Yamazaki, I. Iwata, M. Hisada, S. Shimizu, N. Yamaguchi and Y-C Tung. [Immune Responsiveness of Tumor-Bearing Host and Trial of Modulation] in Japanese, Eng. abs. J. Kanazawa Med. Univ. 10 (Supp.):205-210, 1985.

(98) Yamaguchi, N., S. Shimizu, T. Murayama, T. Saito, R.F. Wang and Y.C. Tong: Immunomodulation by single cellular algae (Chlorella pyrenoidosa) and anti-tumor activities for tumor-bearing mice. Presented at the Third International Congress of Developmental and Comparative Immunology, Reims, France, July 7-13, 1985.

(99) Konishi, F., K. Tanaka, K. Himeno, K. Taniguchi, and K. Nomoto: Antitumor effect induced by a hot water extract of Chlorella vulgaris (CE): Resistance to Meth-A tumor growth mediated by CE-induced polymorphonuclear leukocytes. Cancer Immunol. Immunother (1985) 19:73-78.

(100) Hixson, J.R.: Beta-Carotene Showing Promise as Topical Agent. Medical Tribune August 6, 1986, p 3.

(101) Umezawa, I., K. Komiyama, N. Shibukawa, M. Mori and Y. Kojima: An Acidic Polysaccharide, chlon A, from Chlorella pyrenoidosa. Chemotherapy 30(9), 1041-1045, 1982.

(102) Nathan, DF, Murray, HW, Wiebe, ME, Rubin, BY. Identification of Interferon-gamma as the lymphokine that activates human macrophage oxidative metabolism and antimicrobial activity. J.Exp. Med. 1983; 158:670-89.

(103) Nathan, CF. E. Al. Local and Systemic Effects of Intradermal Recombinant Interferon-Gamma in Patients with Lepromatous Leprosy. N. Engl. J.Med. 1986; 315-15.

(104) Murray, H.W., Spitalny, G.L., Nathan, C.F.:Activation of mouse peritoneal macrophages in vitro and in vivo by interferon-gamma. J. Immunol. 1985;134:1619-22.
(105) References given in:Nathan, CF. et. al. Local and Systemic Effects of Intradermal Recombinant Interferon-Gamma in Patients with Lepromatous Leprosy. N. Engl. J. Med. 1986; 315-15
(106) Umezawa, I., K. Komiyama, N. Shibukawa, M. Mori, and Y. Kojima: An acidic polysaccharide, chlon A, from Chlorella pyrenoidosa. Chemotherapy 30 (9), 1041-1045, Sept. 1982. (in Jpn).
(107) Brinckerhoff, C.E. et. al. Effect of retinoids on rheumatoid arthritis, a proliferative and invasive non-malignant disease. Ciba Found. Symp. 1985, 113, p. 191-211.
(108) Obert, H.J., Hofschneider, P.H.: Interferon bei chronischer Polyarthritis. Positive Wirkung in der klinischen Prufung. Dtsch. Med. Wochenschr. 1985 Nove 15. 110(46). p. 1766-9.
(109) Waslien, C.I., Calloway, D.H., Margen, S. and Costa, F.:Uric acid levels in men fed algae and yeast as protein sources. J. Food Sci., 1970, 35, 294-298.
(110) Shirota Minoru, et. al. (Regarding the anti-virus components extracted from Chlorella.) Showa 42 nen Nihon nogika gakkai koen yori, 1967.
(111) Murayama, T., Leng-Fang, Wang, N. Yamaguchi, and Yin-Chin Tung: Effect of Various Products Derived from Chlorella Pyrenoidosa Cells on Defense Mechanism of Organism (Immunological Resistance.) The 21st Japan Bacteriology Convention, November, 1984. Personal communication from Tsugiya Murayama, Ph.D., Dept. of Microbiology, Kanazawa Medical University, Uchinada-Machi, Kahoku-Gun, Ishikawa-Ken, 920-02 Japan.
(112) Fukada, T., M. Hoshino, H. Endo, M. Mukai and M. Shirota:Photodynamic Antiviral Substance Extracted from Chlorella Cells. Applied Microbiology 16(11) p.1809-1810, 1968.
(113) Kashiwa, Y. and Y. Tanaka: Effect of Chlorella on the changes in the body weight and rate of catching cold of the 1966 training fleet crew. Reported at the Japan Medical Science Meeting, Nagoya, Japan 1966.
(114) White, R.C., G.A. Barber: An acidic polysaccharide from the cell wall of Chlorella pyrenoidosa. Biochem. Biophy. Acta 264, 117-128, 1972.
(115) Figge, F.H.J., et. al. Cancer detection and therapy. Affinity of neoplastic, embryonic, and traumatized tissues for porphyrins and metalloporphyrins. Proc. Soc. Exp. Biol. & Med. 68:640, 1948.
(116) Tixier, L., et. al. La therapeutique des maladies par exces de cholesterol. Rev. Med. Paris 54:201, 1937.
(117) Dupont, R. et Duhamel, G., Chlorophyll et cancer. Bull.
(118) Doiron, D.R. and C.J. Gomer. Porphyrin Localization and Treatment of Tumors. Introduction. Alan R. Liss., New York 1984. Assoc. Franc. l'etude cancer. 24:15, 1935.

(119) Vermeil, C. and O. Morin. Role experimental des algues unicellulaires Prototheca et Chlorella (Chlorellaceae) dans l'immunogenese anticancereuse (sarcome murin BP 8). Societe de Biologie de Rennes, Seance du 21, Avril 1976.
(120) Neveu, P.J., O. Morin, M. Miegeville, B.P. LeMevel and C. Vermeil:Modulation of antibody synthesis by an anti-tumor alga. Experientia 34/12, 1644-1645, 15.12.78
(121) Umezawa, I., K. Komiyama, N. Shibukawa, M. Mori, and Y. Kojima: An acidic polysaccharide, chlon A, from Chlorella pyrenoidosa. Chemotherapy 30 (9), 1041-1045, Sept. 1982. (in Jpn).
(122) Takechi, Yoshiro, "Chlorella—Its Basis and Application. Publ. by Gakushu Kenku-Sha, Tokyo, Japan. Nov. 30, 1971 p. 53.
(123) Kollman, H.V. and R. Schmidt. Algae, The Modern Manna? Lets Live, December 1978.
(124) Matsueda, S., J. Ichita, K. Abe, H. Karasawa, K. Shinpo. Studies on antitumor active glycoprotein from Chlorella vulgaris. Yajugaku-Zasshi 102:447-451, May, 1982.
(125) Nomoto, K., T. Yokokura, H. Satoh, and K. Mutai. Anti-tumor effect by oral administration of Chlorella extract. PCM-4. Gan-To-Kagaku-Ryoho, 10(3), 781-5, March, 1983 (in Jpn).
(126) Tanaka, K., F. Konishi, K. Himeno, K. Taniguchi, and K. Nomoto. Augmentation of antitumor resistance by a strain of unicellular green algae, Chlorella vulgaris. Cancer Immunol. Immunother. 17:90-94, 1984.
(127) Yamaguchi, N. S. Shimizu, T. Murayama, T. Saito, R.F. Wang and Y.C. Tong: Immunomodulation by single cellular algae (Chlorella pyrenoidosa) and anti-tumor activities for tumor-bearing mice. Presented at the Third International Congress of Developmental and Comparative Immunology, Reims, France, July 7-13, 1985.
(128) Vermeil, C., Morin, O., Le-Bodic, L. [The stimulation of tumoricidal peritoneal macrophages can be directly induced by peritoneal implantation of unicellular algae in humans]. Arch-Inst-Pasteur-Tunis. Mar-Jun. 62 (1-2), p. 91-4, 1985.
(129) Tryggvason, K.: Pattern of basement membrane degradation by metastatic tumor cell enzymes. p. 151-166, Chapter 13. In:Biochemistry and Molecular Genetics of Cancer Metastasis. Eds. K. Lapis, L.A. Liotta and A.S. Rabson. Martinus Nijhoff Publishing 1985.
(130) Hoefer-Janker, H. et. al. Arztl. Praxis 23, 538, 542, 1969.
(131) Hoefer-Janker, H. et. al. Kreosarzt 24, 203, 1969.
(132) Hoefer-Janker, H. et. al. Arztl. Praxis 23, 2805, 1971.
(133) Scheef, W. Clinical Experiences with High Concentrations of Vitamin A in Oncology. 14th International Cancer Congress Budapest, Hungary, August 21-27, 1986.
(134) Ebana, Kyoko. Biological significance of Chlorella polysaccharide. VI, Inhibition of experimental hyperlipemia by a mixture of papain, calcium pantothenate, and Chlorella polysaccharide. Fukushima-Ken Eisei Kenkyusho Kenkyu Hokoku 1969, 17(2), 15-20 (Japan).

(135) Lin, J-K., et. al. Effect of Chlorella on Serum Cholesterol of Rats. Taiwan Medical Science Journal Sept. 1981.

(136) Sakuno, T. et. al. Inhibitory effect of Chlorella on increases in serum and liver cholesterol levels of rats. Health Industry Newsletter March 25, 1978.

(137) Health Industry Newspaper. Summaries of Chlorella related reports made public at the Agricultural Chemical convention. March 25, 1978.

(138) Okuda, M. J. Hasegawa, M. Sonoda, T. Okabe, Y. Tanaka: The effects of Chlorella on the levels of cholesterol in serum and liver. Japanese J. Nutr. 33;3-B, 1975.

(139) Wang, C.-J., S.-J. Shiow and J.K. Lin. Effect of chlorella on the level of serum cholesterol in rats. J. Formosan Med. Assoc. 80:929-933, 1981.

(140) Yoshiro, T.: Chlorella-Its Basis and Application. Publ. by Gakushu Kenku-Sha, Tokyo, Japan Nov. 30, 1971.

(141)

(142) Okamoto, K., Y. Iizuka, T. Murakami, H. Miyake and T. Suziki:Effects of Chlorella Alkali Extract on Blood Pressure in SHR. Jpn. Heart J. 19(4), P.622-623, July, 1978.

(143) Murakami, T. et. al. Effect of Heterotrophy Chlorella on blood pressure and development of apoplexy in hypertensive rats. Food Research Institute, Kinki University. Reported at the Agricultural-Chemical Convention in Sendai, Japan, March 30, 1983.

(144) Shimizu, M., N. Yamada, M. Hisada, J. Suzuki, I. Inata. Effect of Chlorella on Human Pulse Wave Velocity. Kanazawa Medical University, Dept. of Serology. April 8, 1985.

(145) Burgi, E. Chlorophyll als Pharmakon. 1932.

(146) Angelo, L. (Clinical studies on chlorophyll). LaRiforma Medica 47:83. 1931.

(147) Zickgraf, G. Ueber parenterale. Chlorophyll in therapie. Munch. Med. Wochenschrift 79:998. 1932.

(148) Hasuda, Shioichi and Y. Mito. Experience in taking chlorella for healing the less curable wound. Chapter V. Application of Chlorella on Medicine and Food. Taiwan Chlorella Manufacture Co., Ltd., P.O. Box 1250 Taipei, Taiwan, 71, sec. 2 Nanking East Road, Taipi, Taiwan R.O.C.

(149) Hasuda, Shioichi and Y. Mito. Medical Examinations of chlorella for hard to cure wounds. Medical Examinations and New Drugs 3,3, 1966. Japan.

(150) Jorgensen, Jorgen and Jacinto Convit. Cultivation of Complexes of Algae with other fresh-water Microorganisms in the Tropics. Chapter 14 pp. 190-196 in: Burlew, John S.:Algal Culture: From Laboratory to Pilot Plant. Carnegie Institution of Washington Publication 600, Washington, D.C. 1953.

(151) Terziev, V., B. Planski, and Y. Encheva. The Effect of chlorella as a prophylactic means against bronchopneumonia. Veterinary Science XX(1), 36-39, 1983.

(152) Pratt, R.: Influence of the size of the inoculum on the growth of Chlorella vulgaris in freshly prepared culture medium. American Journal of Botony, 27:52, 1940.

(153) Pratt, R.:Studies on Chlorella vulgaris, V. Some properties of photosynthesis by a growth inhibitor formed by Chlorella cells. American J. of Botony, 29:142, 1942.

(154) Pratt, R.:Studies of Chlorella vulgaris, VI. Retardation of photosynthesis by a growth inhibiting substance from Chlorella vulgaris. American J. of Botony 30:32, 1943.

(155) Pratt, R. and Fong, J. Studies on Chlorella vulgaris, II. Further evidence that Chlorella cells form inhibiting substance. American J. of Botony, 27:431, 1940.

(156) Pratt, R. Daniels, T.C., Eiler, J.J., Gunnison, J.B., Kumler, W.D., Oneto, J.F., Strait, L.A., Spoehr, H.A., Harden, G.J.: Chlorellin, an antibacterial substance from Chlorella. Science 99:351-352, 1944.

(157) Pratt, R., and Mautner, H.: Antibiotic activity of seaweed extracts. J. of American Pharmaceutical Association (Scientific edition), 40:575, 1951.

(158) Krishnamurty, Goteti Bala. The Effect of Algae on Selected Bacterial Populations in Sewage. M.S. Thesis. UCLA. 1959.

(159) Yamaguchi, Y. Toikawa, M. Suzuki, R., Hara, T., Warita, Y. Therapy for peptic ulcers by chlorella. Nippon Iji Shinpo No. 1997, pp25-27, 1962. (Japanese)

(160) Taiwan Chlorella Manufacture Co., Ltd.:Application of Chlorella on Medicine and Food. Report Summaries. P.O. Box: 1250 Taipei, Taiwan, 71, Sec. 2 Nanking East Road, Taipi, Taiwan R.O.C.

(161) Fukui, S.: The studies of nutrition controls to long term prison inmates. 14th. Orthosis Medical Society, 1967.

(162) Fukui. S. and Sasaki, T.: About the efficacy of chlorella pills. Medical examinations and new drugs. 6, 11, 1969. Japan.

(163) Asanov, P.L. (Content of protein and its fractions in the blood serum of piglets given supplements of chlorella) Uzb. biol. Z. 1969, No. 6, 51-52. Russian.

(164) Tamiya, Hiroshi. Role of Algae as Food. Proceedings of the Symposium on Algology, New Delhi, 1959. pp379-389.

(165) Hunter, B. and P. Batham, Toxocologists. Huntington Research Centre toxicology report, November 1972. Paper submitted to the FDA for approval of Chlorella as a food substance.

(166) Surrey and Morse. FDA Petition for the importation of chlorella.

(167) Hunter, Brian and Peter Batham, "Acute Oral Toxicity to Rats of Green Chlorella and Yellow Chlorella Powders," Huntington Research Centre, Huntington, England (November 1972).

(168) Dam, R., Lee, S., Fry, P.C., and Fox, H.: Utilization of algae as a protein source for humans. J. Nutrition, 1965, 86, 376-382.

(169) Sonoda, M. (Effect of chlorella extract on pregnancy anemia) Jap. J. Nutr. (1972) 30(5), 218-225.

(170) Basu, T.K., U. Chan, A. Fields. Vitamin A (Retinol) and Epithelial Cancer in Man. Prasad(ed.), Vitamins, Nutrition, and Cancer, pp. 33-45 (Karger, Basel 1984)

(171) Schumacher, H.R. Jr. Ultrastructure of the synovial membrane. Ann. Clin. Lab. Sci. 5(6), pp. 489-98, 1975.

(172) Lorch, D. and A. Weber. Accumulation, Toxicity and Localization of Lead in cryptograms: Experimental Results. In: Heavy Metals in Water Organisms. Symposia Biologica Hungarica, Akademiai Kiado, Budapest. 29, pp51-60, 1985.

(173) Bogorad, L., and Granick, S. (1953). J. Biol. Chem. 202, 793.

(174) Granick, S., Bogorad, L., and Jaffe, H. (1953) J.Biol. Chem. 202, 801.

(175) Jitsukawa, K., R. Suizu and A. Hidano. Chlorella Photosensitization. International J. Derm. 2(4), 263-268, May 1984.

(176) Kimura, S., T. Isobe, H. Sai, Y. Takahashi. The role of lipid peroxidation on the development of photosensitivity syndrome by pheophorbide a. In: Lipid Peroxides in Biology and Medicine. ed. Kunio Yagi. pp. 243-254. Academic Press 1982.

(177) Kimura, S., and Y. Takahashi. Preventive Effects of L-Ascorbic Acid and Calcium Pantothenate against Photosensitive Actions Induced by Pheophorbide a and Hematoporphyrin. J. Nutr. Sci. Vitaminol. 27, 521-527, 1981.

(178) Takeda, Y., Y. Saito and M. Uchiyama. Determination of pheophorbide a, pyropheophorbide a and phytol. J. Chromatography 280, 188-193, 1983.

(179) Kimura, S. The Revelation of Toxicity Which is Caused by some Poisonous Substances Derived from Foodstuffs and its modification under Nutritional Conditions. Yakugaku Zasshi 104(5), 423-439, 1984. (Jpn).

(180) Hashimoto, K. High Performance Liquid Chromatographic Determination of Pheophorbide in Chlorella Tablets. Yakugaku Zasshi 105(1), 33-37, 1985. (Jpn).

(181) Miki, K., O. Tajima, E. Matsuda, K. Yamada, and T. Fukimbara. Isolation and Identification of a Photodynamic Agent of Chlorella. Nippon Nogeikagaku Kaishi vol. 54, No. 9, pp. 721-726, 1980. (Jpn).

(182) Komai, Y., Onuki, K., Yamagishi, H. et. al: Photosensitive dermatitis due to chlorella tablets. Food Clit Res 28:747, 1978.

(183) Tamura, Y., Nishigaki, S., Maki, T., et. al: Photosensitive dermatitis caused by chlorella tablets. Ann Rep Tokyo Metr. Res. Lab P.H. 29:250, 1978.

(184) Nagano, T., Y. Watanabe, T. Honma, Y. Suketa, and T. Yamamoto. Absorption and Excretion of Cadmium by the Rat administered Cadmium-containing Chlorella. Eisei Kagaku 24(4), 182-186, 1978.

(185) Hundley, R.F., J.R. Spears and R. Weinstein. Photodynamic cytolosis of arterial smooth muscle cells in vitro: implications for laser angioplasty. JACC 5 (2), 408, Feb. 1985. Abstract.

WHAT'S AN OSTEOPATHIC PHYSICIAN?

If you're like most people, you're not quite sure what a D.O. — a Doctor of Osteopathy — really is.

You may even think an osteopathic physician is someone to see only when you have a problem with your bones or your back.

That's a common mistake. Actually, D.O.'s are fully trained and licensed to practice **all** phases of medicine in all 50 states. They are complete doctors who offer their patients something extra.

Osteopathic physicians perform surgery, deliver babies, treat patients and prescribe medicine in hospitals and offices across the country, and in all branches of the armed services. And these D.O. general practitioners, surgeons and other specialists use all the tools of modern medicine to detect and treat disease.

But they also do more. They are specially trained to perform osteopathic manipulation. That's a technique in which osteopathic physicians use their hands to diagnose illness and treat patients.

They pay particular attention to your joints, bones, muscles and nerves. As a result of manipulation, your circulation is often improved. And a normal blood and nerve supply help your body to heal itself.

Osteopathic physicians treat patients in a special way, too. They look at the whole person, not just the part that's sick, such as your arm or leg. They are concerned about **all** of you.

They know that what happens in one part of your body affects other parts, too. That's why most D.O.'s are family doctors. They care for the whole person.

So, now you know. D.O.'s are complete doctors who offer their patients something extra. And who treat them in a special way.